TRIGONOMETRY DEMYSTIFIED

TRIGONOMETRY DEMYSTIFIED

STAN GIBILISCO

McGRAW-HILL
New York   Chicago   San Francisco   Lisbon   London
Madrid   Mexico City   Milan   New Delhi   San Juan
Seoul   Singapore   Sydney   Toronto

The McGraw-Hill Companies

Cataloging-in-Publication Data is on file with the Library of Congress

Copyright © 2003 by The McGraw-Hill Companies, Inc. All rights reserved. Printed in the United States of America. Except as permitted under the United States Copyright Act of 1976, no part of this publication may be reproduced or distributed in any form or by any means, or stored in a data base or retrieval system, without the prior written permission of the publisher.

1 2 3 4 5 6 7 8 9 0   DOC/DOC   0 9 8 7 6 5 4 3

ISBN 0-07-141631-5

The sponsoring editor for this book was Scott L. Grillo and the production supervisor was Pamela A. Pelton. It was set in Times Roman by Keyword Publishing Services Ltd.

Printed and bound by RR Donnelley.

 This book was printed on recycled, acid-free paper containing a minimum of 50% recycled, de-inked fiber.

McGraw-Hill books are available at special quantity discounts to use as premiums and sales promotions, or for use in corporate training programs. For more information, please write to the Director of Special Sales, McGraw-Hill Professional, Two Penn Plaza, New York, NY 10121-2298. Or contact your local bookstore.

CONTENTS

CONTENTS

PREFACE

This book is for people who want to get acquainted with the concepts of basic trigonometry without taking a formal course. It can serve as a supplemental text in a classroom, tutored, or home-schooling environment. It should also be useful for career changers who need to refresh their knowledge of the subject. I recommend that you start at the beginning of this book and go straight through.

This is not a rigorous course in theoretical trigonometry. Such a course defines *postulates* (or *axioms*) and provides deductive proofs of statements called *theorems* by applying mathematical logic. Proofs are generally omitted in this book for the sake of simplicity and clarity. Emphasis here is on practical aspects and scientific applications. You should have knowledge of middle-school algebra before you begin this book.

This introductory work contains an abundance of practice quiz, test, and exam questions. They are all multiple-choice, and are similar to the sorts of questions used in standardized tests. There is a short quiz at the end of every chapter. The quizzes are "open-book." You may (and should) refer to the chapter texts when taking them. When you think you're ready, take the quiz, write down your answers, and then give your list of answers to a friend. Have the friend tell you your score, but not which questions you got wrong. The answers are listed in the back of the book. Stick with a chapter until you get most of the answers correct.

This book is divided into two sections. At the end of each section is a multiple-choice test. Take these tests when you're done with the respective sections and have taken all the chapter quizzes. The section tests are "closed-book," but the questions are not as difficult as those in the quizzes. A satisfactory score is three-quarters of the answers correct. Again, answers are in the back of the book.

There is a final exam at the end of this course. It contains questions drawn uniformly from all the chapters in the book. Take it when you have finished both sections, both section tests, and all of the chapter quizzes. A satisfactory score is at least 75 percent correct answers.

With the section tests and the final exam, as with the quizzes, have a friend tell you your score without letting you know which questions you missed. That way, you will not subconsciously memorize the answers. You can check to see where your knowledge is strong and where it is not.

I recommend that you complete one chapter a week. An hour or two daily ought to be enough time for this. When you're done with the course, you can use this book, with its comprehensive index, as a permanent reference.

Suggestions for future editions are welcome.

STAN GIBILISCO

ACKNOWLEDGMENTS

Illustrations in this book were generated with *CorelDRAW*. Some clip art is courtesy of Corel Corporation, 1600 Carling Avenue, Ottawa, Ontario, Canada K1Z 8R7.

I extend thanks to Emma Previato of Boston University, who helped with the technical editing of the manuscript for this book.

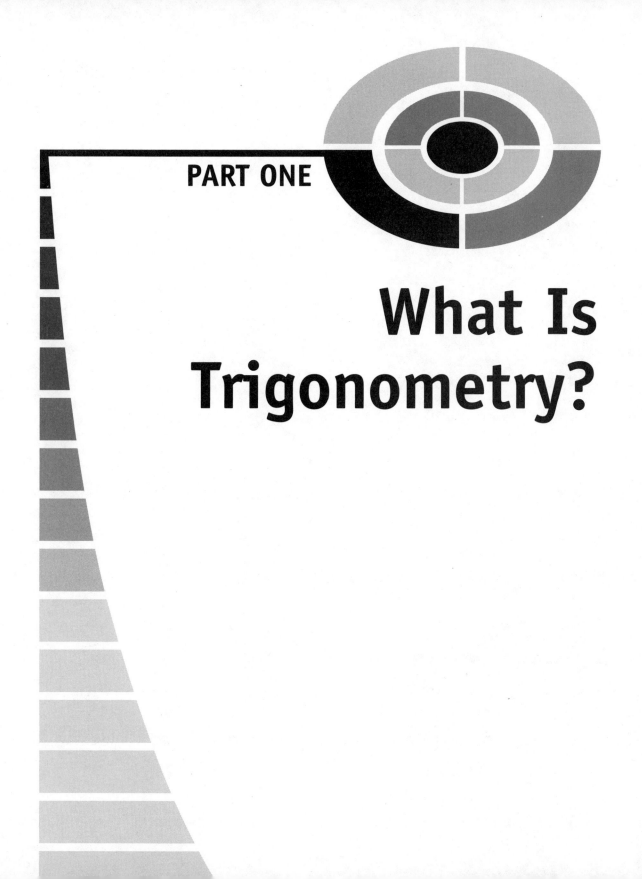

PART ONE

What Is Trigonometry?

The Circle Model

Trigonometry involves angles and their relationships to distances. All of these relationships arise from the characteristics of a circle, and can be defined on the basis of the graph of a circle in the *Cartesian plane*.

The Cartesian Plane

The Cartesian plane, also called the *rectangular coordinate plane* or *rectangular coordinates*, consists of two number lines that intersect at a right angle. This makes it possible to graph equations that relate one variable to another. Most such graphs look like lines or curves.

TWO PERPENDICULAR NUMBER LINES

Figure 1-1 illustrates the simplest possible set of rectangular coordinates. Both number lines have uniform increments. That is, the points on the axes that represent consecutive integers are always the same distance apart. The two number lines intersect at their zero points. The horizontal (or east/west) axis is called the *x axis*; the vertical (or north/south) axis is called the *y axis*.

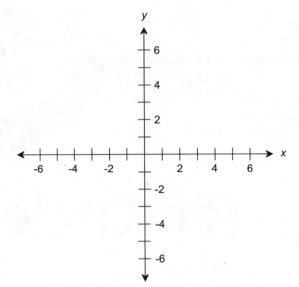

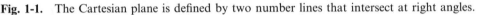

Fig. 1-1.   The Cartesian plane is defined by two number lines that intersect at right angles.

ORDERED PAIRS

Figure 1-2 shows two points plotted in rectangular coordinates. Points are denoted as *ordered pairs* in the form (*x*,*y*) in which the first number represents the value on the *x* axis and the second number represents the value on the *y* axis. The word "ordered" means that the order in which the numbers are listed is important. For example, the ordered pair (3.5,5.0) is not the same as the ordered pair (5.0,3.5), even though both pairs contain the same two numbers.

In ordered-pair notation, there is no space after the comma, as there is in the notation of a set or sequence. When denoting an ordered pair, it is customary to place the two numbers or variables together right up against the comma.

ABSCISSA AND ORDINATE

In most sets of coordinates where the axes are labeled *x* and *y*, the variable *y* is called the *dependent variable* (because its value "depends" on the value of *x*), and the variable *x* is called the *independent variable*. The independent-variable coordinate (usually *x*) of a point on the Cartesian plane is called the *abscissa*, and the dependent-variable coordinate (usually *y*) is called the *ordinate*. The point (0,0) is called the *origin*.

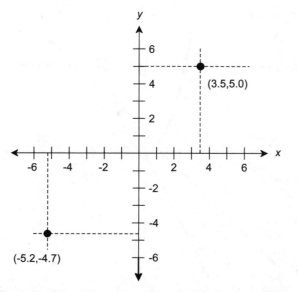

Fig. 1-2.   Two points, plotted in rectangular coordinates.

In Fig. 1-2, two points are shown, one with an abscissa of 3.5 and an ordinate of 5.0, and the other with an abscissa of −5.2 and an ordinate of −4.7.

RELATIONS

Mathematical relationships, technically called *relations*, between two variables x and y can be written in such a way that y is expressed in terms of x. The following are some examples of relations denoted in this form:

$$y = 5$$
$$y = x + 1$$
$$y = 2x$$
$$y = x^2$$

SOME SIMPLE GRAPHS

Figure 1-3 shows how the graphs of the above equations look on the Cartesian plane. Mathematicians and scientists call such graphs *curves*, even if they are straight lines.

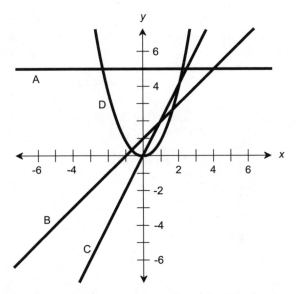

Fig. 1-3.   Graphs of four simple functions. See text for details.

The graph of $y = 5$ (curve A) is a horizontal line passing through the point (0,5) on the y axis. The graph of $y = x + 1$ (curve B) is a straight line that ramps upward at a 45° angle (from left to right) and passes through the point (0,1) on the y axis. The graph of $y = 2x$ (curve C) is a straight line that ramps upward more steeply, and that passes through the origin. The graph of $y = x^2$ (curve D) is known as a *parabola*. In this case the parabola rests on the origin, opens upward, and is symmetrical with respect to the y axis.

RELATIONS VS FUNCTIONS

All of the relations graphed in Fig. 1-3 have something in common. For every abscissa, each relation contains at most one ordinate. Never does a curve portray two or more ordinates for a single abscissa, although one of them (the parabola, curve D) has two abscissas for all positive ordinates.

A mathematical relation in which every abscissa corresponds to at most one ordinate is called a *function*. All of the curves shown in Fig. 1-3 are graphs of functions of y in terms of x. In addition, curves A, B, and C show functions of x in terms of y (if we want to "go non-standard" and consider y as the independent variable and x as the dependent variable).

Curve D does not represent a function of x in terms of y. If x is considered the dependent variable, then there are some values of y (that is, some abscissas) for which there exist two values of x (ordinates).

PROBLEM 1-1
Suppose a certain relation has a graph that looks like a circle. Is this a function of *y* in terms of *x*? Is it a function of *x* in terms of *y*?

SOLUTION 1-1
The answer is no in both cases. Figure 1-4 shows why. A simple visual "test" to determine whether or not a given relation is a function is to imagine an infinitely long, straight line parallel to the dependent-variable axis, and that can be moved back and forth. If the curve ever intersects the line at more than one point, then the curve is not a function.

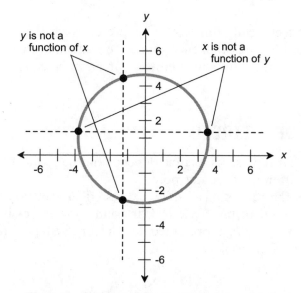

Fig. 1-4.   Illustration for Problems 1-1 and 1-2.

A "vertical line" (parallel to the *y* axis) test can be used to determine whether or not the circle is a function of the form $y = f(x)$, meaning "*y* is a function of *x*." Obviously, the answer is no, because there are some positions of the line for which the line intersects the circle at two points.

A "horizontal line" (parallel to the *x* axis) test can be used to determine if the circle is a function of the form $x = f(y)$, meaning "*x* is a function of *y*." Again the answer is no; there are some positions of the line for which the line intersects the circle twice.

PROBLEM 1-2
How could the circle as shown in Fig. 1-4 be modified to become a function of *y* in terms of *x*?

SOLUTION 1-2

Part of the circle must be removed, such that the resulting curve passes the "vertical line" test. For example, either the upper or the lower semicircle can be taken away, and the resulting graph will denote y as a function of x. But these are not the only ways to modify the circle to get a graph of a function. There are infinitely many ways in which the circle can be partially removed or broken up in order to get a graph of a function. Use your imagination!

Circles in the Plane

Circles are not technically functions as represented in the Cartesian coordinate system, but they are often encountered in mathematics and science. They are defined by equations in which either x or y can be considered the dependent variable.

EQUATION OF A CIRCLE

The equation that represents a circle depends on the radius of the circle, and also on the location of its center point.

Suppose r is the radius of a circle, expressed in arbitrary units. Imagine that the center point of the circle in Cartesian coordinates is located at the point $x = a$ and $y = b$, represented by the ordered pair (a,b). Then the equation of that circle looks like this:

$$(x - a)^2 + (y - b)^2 = r^2$$

If the center of the circle happens to be at the origin, that is, at $(0,0)$ on the coordinate plane, then the general equation is simpler:

$$x^2 + y^2 = r^2$$

THE UNIT CIRCLE

Consider a circle in rectangular coordinates with the following equation:

$$x^2 + y^2 = 1$$

This is called the *unit circle* because its radius is one unit, and it is centered at the origin $(0,0)$. This circle is significant, because it gives us a simple basis to define the common trigonometric functions, which are called *circular functions*. We'll define these shortly.

IT'S GREEK TO US

In geometry, and especially in trigonometry, mathematicians and scientists have acquired the habit of using Greek letters to represent angles. The most common symbol for this purpose is an italicized, lowercase Greek theta (pronounced "THAY-tuh"). It looks like a numeral zero leaning to the right, with a horizontal line through it (θ).

When writing about two different angles, a second Greek letter is used along with θ. Most often, it is the italicized, lowercase letter phi (pronounced "fie" or "fee"). It looks like a lowercase English letter o leaning to the right, with a forward slash through it (ϕ). You should get used to these symbols, because if you have anything to do with engineering and science, you're going to find them often.

Sometimes the italic, lowercase Greek alpha ("AL-fuh"), beta ("BAY-tuh"), and gamma ("GAM-uh") are used to represent angles. These, respectively, look like this: α, β, γ. When things get messy and there are a lot of angles to talk about, numeric subscripts are sometimes used with Greek letters, so don't be surprised if you see angles denoted θ_1, θ_2, θ_3, and so on.

RADIANS

Imagine two rays emanating outward from the center point of a circle. The rays each intersect the circle at a point. Call these points P and Q. Suppose the distance between P and Q, as measured along the arc of the circle, is equal to the radius of the circle. Then the measure of the angle between the rays is one *radian* (1 rad).

There are 2π rad in a full circle, where π (the lowercase, non-italic Greek letter pi, pronounced "pie") stands for the ratio of a circle's circumference to its diameter. The value of π is approximately 3.14159265359, often rounded off to 3.14159 or 3.14. A quarter circle is $\pi/2$ rad, a half circle is π rad, and a three-quarter circle is $3\pi/2$ rad. Mathematicians generally prefer the radian when working with trigonometric functions, and the "rad" is left out. So if you see something like $\theta_1 = \pi/4$, you know the angle θ_1 is expressed in radians.

DEGREES, MINUTES, SECONDS

The angular *degree* (°), also called the *degree of arc*, is the unit of angular measure most familiar to lay people. One degree (1°) is 1/360 of a full circle. An angle of 90° represents a quarter circle, 180° represents a half circle, 270° represents a three-quarter circle, and 360° represents a full circle. A right angle has a measure of 90°, an acute angle has a measure of more than 0° but less than 90°, and an obtuse angle has a measure of more than 90° but less than 180°.

To denote the measures of tiny angles, or to precisely denote the measures of angles in general, smaller units are used. One *minute of arc* or *arc minute*, symbolized by an apostrophe or accent (′) or abbreviated as m or min, is 1/60 of a degree. One *second of arc* or *arc second*, symbolized by a closing quotation mark (″) or abbreviated as s or sec, is 1/60 of an arc minute or 1/3600 of a degree. An example of an angle in this notation is 30° 15′ 0″, which denotes 30 degrees, 15 minutes, 0 seconds.

Alternatively, fractions of a degree can be denoted in decimal form. You might see, for example, 30.25°. This is the same as 30° 15′ 0″. Decimal fractions of degrees are easier to work with than the minute/second scheme when angles must be added and subtracted, or when using a conventional calculator to work out trigonometry problems. Nevertheless, the minute/second system, like the English system of measurements, remains in widespread use.

PROBLEM 1-3
A text discussion tells you that $\theta_1 = \pi/4$. What is the measure of θ_1 in degrees?

SOLUTION 1-3
There are 2π rad in a full circle of 360°. The value $\pi/4$ is equal to 1/8 of 2π. Therefore, the angle θ_1 is 1/8 of a full circle, or 45°.

PROBLEM 1-4
Suppose your town is listed in an almanac as being at 40° 20′ north latitude and 93° 48′ west longitude. What are these values in decimal form? Express your answers to two decimal places.

SOLUTION 1-4
There are 60 minutes of arc in one degree. To calculate the latitude, note that $20′ = (20/60)° = 0.33°$; that means the latitude is 40.33° north. To calculate the longitude, note that $48′ = (48/60)° = 0.80°$; that means the longitude is 93.80° west.

Primary Circular Functions

Consider a circle in rectangular coordinates with the following equation:

$$x^2 + y^2 = 1$$

This equation, as defined earlier in this chapter, represents the unit circle. Let θ be an angle whose apex is at the origin, and that is measured counterclockwise from the x axis, as shown in Fig. 1-5. Suppose this angle corresponds to a ray that intersects the unit circle at some point $P = (x_0, y_0)$. We can define three basic trigonometric functions, called circular functions, of the angle θ in a simple and elegant way.

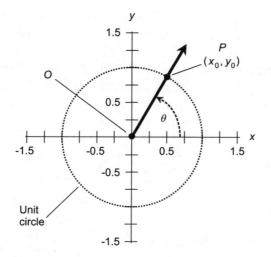

Fig. 1-5.   The unit circle is the basis for the trigonometric functions.

THE SINE FUNCTION

The ray from the origin (point O) passing outward through point P can be called ray OP. Imagine ray OP pointing right along the x axis, and then starting to rotate counterclockwise on its end point O, as if point O is a mechanical bearing. The point P, represented by coordinates (x_0, y_0), therefore revolves around point O, following the perimeter of the unit circle.

Imagine what happens to the value of y_0 (the ordinate of point P) during one complete revolution of ray OP. The ordinate of P starts out at $y_0 = 0$, then increases until it reaches $y_0 = 1$ after P has gone 90° or $\pi/2$ rad around the circle ($\theta = 90° = \pi/2$). After that, y_0 begins to decrease, getting back to y_0 = 0 when P has gone 180° or π rad around the circle ($\theta = 180° = \pi$). As P continues on its counterclockwise trek, y_0 keeps decreasing until, at $\theta = 270°$ = $3\pi/2$, the value of y_0 reaches its minimum of -1. After that, the value of y_0 rises again until, when P has gone completely around the circle, it returns to $y_0 = 0$ for $\theta = 360° = 2\pi$.

The value of y_0 is defined as the *sine* of the angle θ. The sine function is abbreviated sin, so we can state this simple equation:

$$\sin \theta = y_0$$

CIRCULAR MOTION

Suppose you swing a glowing ball around and around at the end of a string, at a rate of one revolution per second. The ball describes a circle in space (Fig. 1-6A). Imagine that you make the ball orbit around your head so it is always at the same level above the ground or the floor; that is, so that it takes a path that lies in a horizontal plane. Suppose you do this in a dark gymnasium. If a friend stands several meters away, with his or her eyes right in the plane of the ball's orbit, what will your friend see?

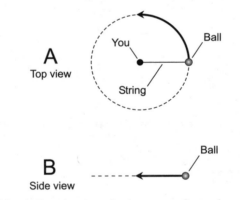

Fig. 1-6. Orbiting ball and string. At A, as seen from above; at B, as seen edge-on.

Close your eyes and use your imagination. You should be able to envision that the ball, seen from a few meters away, will appear to oscillate back and forth in a straight line (Fig. 1-6B). It is an illusion: the glowing dot seems to move toward the right, slow down, then stop and reverse its direction, going back toward the left. It moves faster and faster, then slower again, reaching its left-most point, at which it stops and turns around again. This goes on and on, at the rate of one complete cycle per second, because you are swinging the ball around at one revolution per second.

THE SINE WAVE

If you graph the position of the ball, as seen by your friend, with respect to time, the result is a *sine wave* (Fig. 1-7), which is a graphical plot of a sine function. Some sine waves are "taller" than others (corresponding to a longer string), some are "stretched out" (corresponding to a slower rate of rotation), and some are "squashed" (corresponding to a faster rate of rotation). But the characteristic shape of the wave is the same in every case. When the amplitude and the wavelength are multiplied and divided by the appropriate numbers (or *constants*), any sine wave can be made to fit exactly along the curve of any other sine wave.

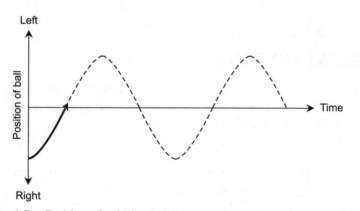

Fig. 1-7.   Position of orbiting ball as seen edge-on, as a function of time.

You can whirl the ball around faster or slower than one revolution per second. The string can be made longer or shorter. These adjustments alter the height and/or the frequency of the sine wave graphed in Fig. 1-7. But the fundamental rule always applies: the sine wave can be reduced to circular motion. Conversely, circular motion in the (x,y) plane can be defined in terms of a general formula:

$$y = a \sin b\theta$$

where a is a constant that depends on the radius of the circle, and b is a constant that depends on the revolution rate.

THE COSINE FUNCTION

Look again at Fig. 1-5. Imagine, once again, a ray from the origin outward through point P on the circle, pointing right along the x axis, and then rotating in a counterclockwise direction.

What happens to the value of x_0 (the abscissa of point P) during one complete revolution of the ray? The abscissa of P starts out at $x_0 = 1$, then decreases until it reaches $x_0 = 0$ when $\theta = 90° = \pi/2$. After that, x_0 continues to decrease, getting down to $x_0 = -1$ when $\theta = 180° = \pi$. As P continues counterclockwise around the circle, x_0 begins to increase again; at $\theta = 270° = 3\pi/2$, the value gets back up to $x_0 = 0$. After that, x_0 increases further until, when P has gone completely around the circle, it returns to $x_0 = 1$ for $\theta = 360° = 2\pi$.

The value of x_0 is defined as the *cosine* of the angle θ. The cosine function is abbreviated cos. So we can write this:

$$\cos\ \theta = x_0$$

THE TANGENT FUNCTION

Once again, refer to Fig. 1-5. The *tangent* (abbreviated tan) of an angle θ is defined using the same ray OP and the same point $P = (x_0, y_0)$ as is done with the sine and cosine functions. The definition is:

$$\tan\ \theta = y_0/x_0$$

Because we already know that $\sin \theta = y_0$ and $\cos \theta = x_0$, we can express the tangent function in terms of the sine and the cosine:

$$\tan\ \theta = \sin\ \theta/\cos\ \theta$$

This function is interesting because, unlike the sine and cosine functions, it "blows up" at certain values of θ. Whenever $x_0 = 0$, the denominator of either quotient above becomes zero. Division by zero is not defined, and that means the tangent function is not defined for any angle θ such that $\cos \theta = 0$. Such angles are all the odd multiples of $90°$ ($\pi/2$ rad).

PROBLEM 1-5
What is tan 45°? Do not perform any calculations. You should be able to infer this without having to write down a single numeral.

SOLUTION 1-5
Draw a diagram of a unit circle, such as the one in Fig. 1-5, and place ray OP such that it subtends an angle of $45°$ with respect to the x axis. That

angle is the angle of which we want to find the tangent. Note that the ray OP also subtends an angle of 45° with respect to the y axis, because the x and y axes are perpendicular (they are oriented at 90° with respect to each other), and 45° is exactly half of 90°. Every point on the ray OP is equally distant from the x and y axes; this includes the point (x_0,y_0). It follows that $x_0 = y_0$, and neither of them is equal to zero. From this, we can conclude that $y_0/x_0 = 1$. According to the definition of the tangent function, therefore, $\tan 45° = 1$.

Secondary Circular Functions

The three functions defined above form the cornerstone for the whole branch of practical mathematics commonly called trigonometry. However, three more circular functions exist. Their values represent the reciprocals of the values of the preceding three functions. To understand the definitions of these functions, look again at Fig. 1-5.

THE COSECANT FUNCTION

Imagine the ray OP, subtending an angle θ with respect to the x axis, and emanating out from the origin and intersecting the unit circle at the point $P = (x_0,y_0)$. The reciprocal of the ordinate, that is, $1/y_0$, is defined as the *cosecant* of the angle θ. The cosecant function is abbreviated csc, so we can state this simple equation:

$$\csc \theta = 1/y_0$$

This function is the reciprocal of the sine function. That is to say, for any angle θ, the following equation is always true as long as $\sin \theta$ is not equal to zero:

$$\csc \theta = 1/(\sin \theta)$$

The cosecant function is not defined for 0° (0 rad), or for any multiple of 180° (π rad). This is because the sine of any such angle is equal to 0, which would mean that the cosecant would have to be equal to $1/0$. But we can't do anything with a quotient in which the denominator is 0. (Resist the temptation to call it "infinity"!)

THE SECANT FUNCTION

Keeping the same vision in mind, consider $1/x_0$. This is defined as the *secant* of the angle θ. The secant function is abbreviated sec, so we can define it like this:

$$\sec \theta = 1/x_0$$

The secant of any angle is the reciprocal of the cosine of that angle. That is, as long as $\cos \theta$ is not equal to zero:

$$\sec \theta = 1/(\cos \theta)$$

The secant function is not defined for $90°$ ($\pi/2$ rad), or for any odd multiple thereof.

THE COTANGENT FUNCTION

There's one more circular function to go. You can guess it by elimination: x_0/y_0. It is called the *cotangent* function, abbreviated cot. For any ray anchored at the origin and crossing the unit circle at an angle θ:

$$\cot \theta = x_0/y_0$$

Because we already know that $\sin \theta = y_0$ and $\cos \theta = x_0$, we can express the cotangent function in terms of the sine and the cosine:

$$\cot \theta = \cos \theta/\sin \theta$$

The cotangent function is the reciprocal of the tangent function:

$$\cot \theta = 1/\tan \theta$$

This function, like the tangent function, "blows up" at certain values of θ. Whenever $y_0 = 0$, the denominator of either quotient above becomes zero, and the cotangent function is not defined. This occurs at all integer multiples of $180°$ (π rad).

CONVENTIONAL ANGLES

Once in a while you will hear or read about an angle whose measure is negative, or whose measure is $360°$ (2π rad) or more. In trigonometry, any such angle can always be reduced to something that is at least $0°$ (0 rad), but less than $360°$ (2π rad). If you look at Fig. 1-5 one more time, you should be able to see why this is true. Even if the ray OP makes more than one complete revolution counterclockwise from the x axis, or if it turns clockwise instead, its orientation can always be defined by some counterclockwise angle of least 0° (0 rad) but less than 360° (2π rad) relative to the x axis.

Any angle φ of the non-standard sort, like 730° or −9π/4 rad, can be reduced to an angle θ that is at least 0° (0 rad) but less than 360° (2π rad) by adding or subtracting some whole-number multiple of 360° (2π rad).

Multiple revolutions of objects, while not usually significant in pure trigonometry, are sometimes important in physics and engineering. We don't have to worry about whether a vector pointing along the positive y axis has undergone 0.25, 1.25, or 101.25 revolutions counterclockwise, or 0.75, 2.75, or 202.75 revolutions clockwise. But scientists must sometimes deal with things like this, and when that happens, non-standard angles such as 36,450° must be expressed in that form.

VALUES OF CIRCULAR FUNCTIONS

Now that you know how the circular functions are defined, you might wonder how the values are calculated. The answer: with an electronic calculator! Most personal computers have a calculator program built into the operating system. You might have to dig around in the operating system folders to find it, but once you do, you can put a shortcut to it on your computer's desktop. Use the calculator in the "scientific" mode.

The values of the sine and cosine function never get smaller than −1 or larger than 1. The values of other functions can vary wildly. Put a few numbers into your calculator and see what happens when you apply the circular functions to them. Pay attention to whether you're using degrees or radians. When the value of a function "blows up" (the denominator in the unit-circle equation defining it becomes zero), you'll get an error message on the calculator.

PROBLEM 1-6
Use a portable scientific calculator, or the calculator program in a personal computer, to find all six circular functions of 66°. Round your answers off to three decimal places. If your calculator does not have keys for the cosecant (csc), secant (sec), or cotangent (cot) functions, first find the sine (sin), cosine (cos), and tangent (tan) respectively, then find the reciprocal, and finally round off your answer to three decimal places.

SOLUTION 1-6
You should get the following results. Be sure your calculator is set to work with degrees, not radians.

$$\sin\ 66° = 0.914$$
$$\cos\ 66° = 0.407$$
$$\tan\ 66° = 2.246$$
$$\csc\ 66° = 1/(\sin\ 66°) = 1.095$$
$$\sec\ 66° = 1/(\cos\ 66°) = 2.459$$
$$\cot\ 66 = 1/(\tan\ 66°) = 0.445$$

Quiz

Refer to the text in this chapter if necessary. A good score is eight correct. Answers are in the back of the book.

1. A relation has the equation $x^2 + y^2 = 16$. The graph of this relation, in Cartesian coordinates, looks like
   (a) a straight line
   (b) a parabola
   (c) a spiral
   (d) a circle

2. The value of tan 90° is
   (a) 0
   (b) 1
   (c) π
   (d) not defined

3. Which of the following statements is true?
   (a) $\tan \theta = 1 \,/\, \cot \theta$, provided $\cot \theta \neq 0$
   (b) $\tan \theta = 1 - \cos \theta$, provided $\cos \theta \neq 0$
   (c) $\tan \theta = 1 + \sin \theta$, provided $\sin \theta \neq 0$
   (d) $\tan \theta + \cot \theta = 0$, provided $\cot \theta \neq 0$ and $\tan \theta \neq 0$

4. With regard to the circular functions, an angle of 5π rad can be considered the same as an angle of
   (a) 0°
   (b) 90°
   (c) 180°
   (d) 270°

5. An ordinate is
   (a) the value of a dependent variable
   (b) the value of an independent variable
   (c) a relation
   (d) a function

6. The sine of 0° is the same as the sine of
   (a) 45°
   (b) 90°
   (c) 180°
   (d) 270°

7. Suppose the tangent of a certain angle is −1.0000, and its cosine is
   −0.7071, approximated to four decimal places. The sine of this angle,
   approximated to four decimal places, is
   (a) 1.0000
   (b) 0.7071
   (c) −0.7071
   (d) 0.0000

8. What is the approximate measure of the angle described in Question 7?
   (a) 0°
   (b) 90°
   (c) 180°
   (d) None of the above

9. Set your scientific calculator, or the calculator program in your computer, to indicate radians. Activate the inverse-function key (in
   Windows, put a checkmark in the box labeled "Inv"). Be sure the
   calculator is set to work with decimal numbers (in Windows, put a
   dot or a check in the space labeled "Dec"). Next, find the difference
   $1 - 2$ using a calculator, so it displays −1. Then hit the "cos" function
   key, thereby finding the measure of the angle, in radians, whose cosine
   is equal to −1. The resulting number on the display is an excellent
   approximation of
   (a) $\pi/2$
   (b) π
   (c) $3\pi/2$
   (d) 2π

10. Use your scientific calculator, or the calculator program in your com-
    puter, to find the cosine of 1.6π rad (that is, $8\pi/5$ rad). Rounding the
    answer to two decimal places, you should get

(a) an error or an extremely large number
(b) 0.95
(c) −0.31
(d) 0.31

A Flurry of Facts

Trigonometry involves countless relationships among lines, angles, and distances. It seems that each situation has its own function or formula. Throw in the Greek symbology, and things can look scary. But all complicated structures are built using simple blocks, and difficult problems can be unraveled (or concocted, if you like) using circular trigonometric functions.

The Right Triangle Model

In the previous chapter, we defined the six circular functions—sine, cosine, tangent, cosecant, secant, and cotangent—in terms of points on a circle. There is another way to define these functions: the *right-triangle model*.

TRIANGLE AND ANGLE NOTATION

In geometry, it is customary to denote triangles by writing an uppercase Greek letter delta ($\triangle$) followed by the names of the three points representing the corners, or *vertices*, of the triangle. For example, if P, Q, and R are the names of three points, then $\triangle PQR$ is the triangle formed by connecting these points with straight line segments. We read this as "triangle PQR."

Angles are denoted by writing the symbol $\angle$ (which resembles an extremely italicized, uppercase English letter L without serifs) followed by the names of three points that uniquely determine the angle. This scheme lets us specify the extent and position of the angle, and also the rotational sense in which it is expressed. For example, if there are three points P, Q, and R, then $\angle PQR$ (read "angle PQR") has the same measure as $\angle RQP$, but in the opposite direction. The middle point, Q in either case, is the *vertex* of the angle.

The rotational sense in which an angle is measured can be significant in physics, astronomy, and engineering, and also when working in coordinate systems. In the Cartesian plane, angles measured counterclockwise are considered positive by convention, while angles measured clockwise are considered negative. If we have $\angle PQR$ that measures $30°$ around a circle in Cartesian coordinates, then $\angle RQP$ measures $-30°$, which is the equivalent of $330°$. The cosines of these two angles happen to be the same, but the sines differ.

RATIOS OF SIDES

Consider a right triangle defined by points P, Q, and R, as shown in Fig. 2-1. Suppose that $\angle QPR$ is a right angle, so $\triangle PQR$ is a *right triangle*. Let d be the length of line segment QP, e be the length of line segment PR, and f be the length of line segment QR. Let θ be $\angle PQR$, the angle measured counterclockwise between line segments QP and QR. The six circular trigonometric functions can be defined as ratios between the lengths of the sides, as follows:

$$\sin \theta = e/f$$
$$\cos \theta = d/f$$
$$\tan \theta = e/d$$
$$\csc \theta = f/e$$
$$\sec \theta = f/d$$
$$\cot \theta = d/e$$

The longest side of a right triangle is always opposite the $90°$ angle, and is called the *hypotenuse*. In Fig. 2-1, this is the side QR whose length is f. The other two sides are called *adjacent sides* because they are both adjacent to the right angle.

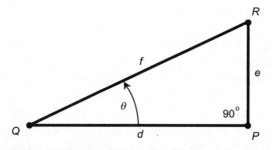

Fig. 2-1.   The right-triangle model for defining trigonometric functions. All right triangles obey the theorem of Pythagoras.

SUM OF ANGLE MEASURES

In any triangle, the sum of the measures of the interior angles is $180°$ (π rad). This holds true whether it is a right triangle or not, as long as all the angles are measured in the plane defined by the three vertices of the triangle.

THEOREM OF PYTHAGORAS

Suppose we have a right triangle defined by points P, Q, and R whose sides have lengths d, e, and f as shown in Fig. 2-1. Then the following equation is always true:

$$d^2 + e^2 = f^2$$

The converse of this is also true: If there is a triangle whose sides have lengths d, e, and f, and the above equation is true, then that triangle is a right triangle. This is known as the *theorem of Pythagoras* (named after the mathematician who supposedly first discovered it, thousands of years ago). It is also called the *Pythagorean theorem*.

  If you want to avoid symbology, you can state the Pythagorean theorem like this: "The square of the length of the hypotenuse of any right triangle is equal to the sum of the squares of the lengths of the other two sides." There's one important condition, however. This holds true only in *Euclidean geometry*, when the triangle is defined in a perfectly "flat" plane. It does not hold for triangles on, say, the surface of a sphere. We are dealing only with Euclidean geometry now, and will not concern ourselves with the idiosyncrasies of *non-Euclidean* situations. That little extra bit of fun is reserved for the last chapter in this book.

RANGE OF ANGLES

In the right-triangle model, the values of the circular functions are defined only for angles between (but in some cases not including) 0° and 90° (0 rad and $\pi/2$ rad). All angles outside this range are better dealt with using the unit-circle model. This is the main shortcoming of the right-triangle model. In the olden days, trigonometry was often taught using the triangle model first, perhaps for the benefit of people who did not understand graphs. But nowadays, when graphs appear on web sites from St. Paul to Sydney, most people are familiar with them.

Using the right-triangle scheme, a trigonometric function is not defined whenever the denominator in its "side ratio" (according to the formulas above) is equal to zero. The length of the hypotenuse (side f) is never zero, but if a right triangle is "squashed" or "squeezed" flat either horizontally or vertically, then the length of one of the adjacent sides (d or e) can become zero. Such objects aren't triangles in the strict sense, because they have only two vertices rather than three—two of the vertex points merge into one—but some people like to include them, in order to take into account angles of 0° (0 rad) and 90° ($\pi/2$ rad).

Geometric purists insist that the right-triangle model can apply only for true triangles, and therefore only to angles θ such that $0° < \theta < 90°$, excluding the angles 0° and 90°. In this sense, the purist is likely to agree with the real-world scientist, who has little interest in 0° angles or "ratios" that have zero in their denominators.

PROBLEM 2-1
Suppose there is a triangle whose sides are 3, 4, and 5 units, respectively. What is the sine of the angle θ opposite the side that measures 3 units?

SOLUTION 2-1
If we are to use the right-triangle model to solve this problem, we must first be certain that a triangle with sides of 3, 4, and 5 units is a right triangle. Otherwise, the scheme won't work. We can test for this by seeing if the Pythagorean theorem applies. If this triangle is a right triangle, then the side measuring 5 units is the hypotenuse, and we should find that $3^2 + 4^2 = 5^2$. Checking, we see that $3^2 = 9$ and $4^2 = 16$. Therefore, $3^2 + 4^2 = 9 + 16 = 25$, which is equal to 5^2. It's a right triangle, all right!

It helps to draw a picture here, after the fashion of Fig. 2-1. Put the angle θ, which we are analyzing, at lower left (corresponding to the vertex point Q). Label the hypotenuse $f = 5$. Now we must figure out which of the other sides should be called d, and which should be called e. We want to find the sine of the angle opposite the side whose length is 3 units, and this angle, in Fig. 2-1, is opposite side PR, whose length is equal to e. So we set $e = 3$. That leaves us with no other choice for d than to set $d = 4$.

According to the formulas above, the sine of the angle in question is equal to e/f. In this case, that means $\sin \theta = 3/5 = 0.6$.

PROBLEM 2-2
What are the values of the other five circular functions for the angle θ as defined in Problem 2-1?

SOLUTION 2-2
Simply plug numbers into the formulas given above, representing the ratios of the lengths of sides in the right triangle:

$$\cos \theta = d/f = 4/5 = 0.8$$
$$\tan \theta = e/d = 3/4 = 0.75$$
$$\csc \theta = f/e = 5/3 \approx 1.67$$
$$\sec \theta = f/d = 5/4 = 1.25$$
$$\cot \theta = d/e = 4/3 \approx 1.33$$

SQUIGGLY OR STRAIGHT?

You will notice a new symbol in the above solution: the squiggly equals sign ($\approx$). This reads "is approximately equal to." It is used by some scientists and mathematicians when working with decimal numbers that are approximations of the actual numerical values. It is also used when instrument readings are known to contain some error.

There is a lot of carelessness when it comes to the use of the squiggly equals sign. The straight equals sign ($=$) is often used even when, if one is to be rigorous, the squiggly equals sign ought to be used. But the reverse situation is not encountered in pure mathematics. You will never see a mathematician seriously write, for example, $5 + 3 \approx 8$, although a technician might be able to get away with it if the numbers are based on instrument readings. Henceforth, we won't concern ourselves with the occasionally strained relationship between these two symbols. We will use the straight equals sign throughout, even when stating approximations or rounded-off values.

Pythagorean Extras

The theorem of Pythagoras can be extended to cover two important facts involving the circular trigonometric functions. These are worth remembering.

PYTHAGOREAN THEOREM FOR SINE AND COSINE

The sum of the squares of the sine and cosine of an angle is always equal to 1. The following formula holds:

$$\sin^2 \theta + \cos^2 \theta = 1$$

The expression $\sin^2 \theta$ refers to the sine of the angle, squared (not the sine of the square of the angle). That is to say:

$$\sin^2 \theta = (\sin \theta)^2$$

The same holds true for the cosine, tangent, cosecant, secant, cotangent, and for all other similar expressions you will see in the rest of this book.

PYTHAGOREAN THEOREM FOR SECANT AND TANGENT

The difference between the squares of the secant and tangent of an angle is always equal to either 1 or −1. The following formulas apply for all angles except $\theta = 90°$ ($\pi/2$ rad) and $\theta = 270°$ ($3\pi/2$ rad):

$$\sec^2 \theta - \tan^2 \theta = 1$$
$$\tan^2 \theta - \sec^2 \theta = -1$$

USE YOUR CALCULATOR!

Trigonometry is a branch of mathematics with extensive applications. You should not be shy about using a calculator to help solve problems. (Neither should you feel compelled to use a calculator if you can easily solve a problem without one.)

PROBLEM 2-3
Use a drawing of the unit circle to help show why it is true that $\sin^2 \theta + \cos^2 \theta$ = 1 for angles θ greater than 0° and less than 90°. (Hint: a right triangle is involved.)

SOLUTION 2-3

Figure 2-2 shows a drawing of the unit circle, with the angle θ defined counterclockwise between the x axis and a ray emanating from the origin. When the angle is greater than $0°$ but less than $90°$, a right triangle is formed, with a segment of the ray as the hypotenuse. The length of this segment is equal to the radius of the unit circle, and this radius, by definition, is 1 unit. According to the Pythagorean theorem for right triangles, the square of the length of the hypotenuse is equal to the sum of the squares of the lengths of the other two sides. It is easy to see from Fig. 2-2 that the lengths of these other two sides are $\sin \theta$ and $\cos \theta$. Therefore

$$(\sin\ \theta)^2 + (\cos\ \theta)^2 = 1^2$$

which is the same as saying that $\sin^2 \theta + \cos^2 \theta = 1$.

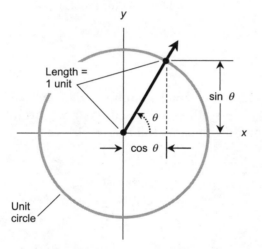

Fig. 2-2.   Illustration for Problem 2-3.

PROBLEM 2-4

Use another drawing of the unit circle to help show why it is true that $\sin^2 \theta + \cos^2 \theta = 1$ for angles θ greater than $270°$ and less than $360°$. (Hint: this range of angles can be thought of as the range between, but not including, $-90°$ and $0°$.)

SOLUTION 2-4

Figure 2-3 shows how this can be done. Draw a mirror image of Fig. 2-2, with the angle θ defined clockwise instead of counterclockwise. Again we have a right triangle; and this triangle, like all right triangles, must obey the Pythagorean theorem.

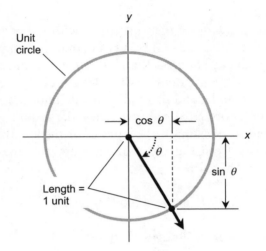

Fig. 2-3.   Illustration for Problem 2-4.

Identities

The following paragraphs depict common *trigonometric identities* for the circular functions. Unless otherwise specified, these formulas apply to angles θ and ϕ in the standard range, as follows:

$$0 \text{ rad } \leq \theta < 2\pi \text{ rad}$$
$$0° \leq \theta < 360°$$
$$0 \text{ rad } \leq \phi < 2\pi \text{ rad}$$
$$\leq \phi < 360°$$

Angles outside the standard range are converted to values within the standard range by adding or subtracting the appropriate multiple of 360° (2π rad). You might occasionally hear of an angle with negative measure or with a measure of more than 360° (2π rad), but this can always be converted to some angle with positive measure that is at least zero but less than 360° (2π rad).

AN ENCOURAGING WORD

When you look at the next few paragraphs and see one equation after another, peppered with Greek symbols, exponents, and parentheses, don't let them intimidate you. All you have to do when working with them is substitute numbers for the angles, and work through the formulas with a calculator. You are not expected to memorize these formulas. They are here for your reference. If you ever need one of these identities, you can refer back to this chapter and look it up!

Trigonometric identities can be useful in solving complicated angle/distance problems in the real world, because they allow the substitution of "clean" expressions for "messy" ones. It's a lot like computer programming. There are many ways to get a computer to perform a specific task, but one scheme is always more efficient than any of the others. Trigonometric identities are intended to help scientists and engineers minimize the number of calculations necessary to get a desired result. This in turn minimizes the opportunity for errors in the calculations. As any scientist knows, the chance that a mistake will be made goes up in proportion to the number of arithmetic computations required to solve a problem.

SINE OF NEGATIVE ANGLE

The sine of the negative of an angle (an angle measured in the direction opposite to the normal direction) is equal to the negative (additive inverse) of the sine of the angle. The following formula holds:

$$\sin -\theta = -\sin \theta$$

COSINE OF NEGATIVE ANGLE

The cosine of the negative of an angle is equal to the cosine of the angle. The following formula holds:

$$\cos -\theta = \cos \theta$$

TANGENT OF NEGATIVE ANGLE

The tangent of the negative of an angle is equal to the negative (additive inverse) of the tangent of the angle. The following formula applies for all angles except $\theta = 90°$ ($\pi/2$ rad) and $\theta = 270°$ ($3\pi/2$ rad):

$$\tan -\theta = -\tan \theta$$

PROBLEM 2-5
Why does the above formula not work when $\theta = 90°$ ($\pi/2$ rad) or $\theta = 270°$ ($3\pi/2$ rad)?

SOLUTION 2-5

The value of the tangent function is not defined for those angles. Remember that the tangent of any angle is equal to the sine divided by the cosine. The cosine of 90° ($\pi/2$ rad) and the cosine of 270° ($3\pi/2$ rad) are both equal to zero. When a quotient has zero in the denominator, that quotient is not defined. This is also the reason for the restrictions on the angle measures in some of the equations that follow.

COSECANT OF NEGATIVE ANGLE

The cosecant of the negative of an angle is equal to the negative (additive inverse) of the cosecant of the angle. The following formula applies for all angles except $\theta = 0°$ (0 rad) and $\theta = 180°$ (π rad):

$$\csc\ -\theta = -\csc\ \theta$$

SECANT OF NEGATIVE ANGLE

The secant of the negative of an angle is equal to the secant of the angle. The following formula applies for all angles except $\theta = 90°$ ($\pi/2$ rad) and $\theta = 270°$ ($3\pi/2$ rad):

$$\sec\ -\theta = \sec\ \theta$$

COTANGENT OF NEGATIVE ANGLE

The cotangent of the negative of an angle is equal to the negative (additive inverse) of the cotangent of the angle. The following formula applies for all angles except $\theta = 0°$ (0 rad) and $\theta = 180°$ (π rad):

$$\cot\ -\theta = -\cot\ \theta$$

SINE OF DOUBLE ANGLE

The sine of twice any given angle is equal to twice the sine of the original angle times the cosine of the original angle:

$$\sin\ 2\theta = 2\sin\ \theta\ \cos\ \theta$$

COSINE OF DOUBLE ANGLE

The cosine of twice any given angle can be found according to either of the following:

$$\cos 2\theta = 1 - (2 \sin^2 \theta)$$
$$\cos 2\theta = (2 \cos^2 \theta) - 1$$

SINE OF ANGULAR SUM

The sine of the sum of two angles θ and ϕ can be found using this formula:

$$\sin (\theta + \phi) = (\sin \theta)(\cos \phi) + (\cos \theta)(\sin \phi)$$

COSINE OF ANGULAR SUM

The cosine of the sum of two angles θ and ϕ can be found using this formula:

$$\cos (\theta + \phi) = (\cos \theta)(\cos \phi) - (\sin \theta)(\sin \phi)$$

SINE OF ANGULAR DIFFERENCE

The sine of the difference between two angles θ and ϕ can be found using this formula:

$$\sin (\theta - \phi) = (\sin \theta)(\cos \phi) - (\cos \theta)(\sin \phi)$$

COSINE OF ANGULAR DIFFERENCE

The cosine of the difference between two angles θ and ϕ can be found using this formula:

$$\cos (\theta - \phi) = (\cos \theta)(\cos \phi) + (\sin \theta)(\sin \phi)$$

That's enough fact-stating for now. Some of these expressions look messy, but they involve nothing more than addition, subtraction, multiplication, division, squaring, and taking the square roots of numbers you work out on a calculator.

PRECEDENCE OF OPERATIONS

When various operations and functions appear in an expression that you want to solve or simplify, there is a well-defined protocol to follow. If you have trouble comprehending the sequence in which operations should be performed, use a pencil and scratch paper to write down the numbers derived by performing functions on variables; then add, subtract, multiply, divide, or whatever, according to the following rules of precedence.

- Simplify all expressions within parentheses from the inside out
- Perform all exponential operations, proceeding from left to right
- Perform all products and quotients, proceeding from left to right
- Perform all sums and differences, proceeding from left to right

Here are a couple of examples of this process, in which the order of the numerals and operations is the same in each case, but the groupings differ.

$$[(2+3)(-3-1)^2]^2$$
$$= [5 \times (-4)^2]^2$$
$$= (5 \times 16)^2$$
$$= 80^2$$
$$= 6400$$

$$\{[2+(3 \times -3)-1]^2\}^2$$
$$= [(2+(-9)-1)^2]^2$$
$$= (-8^2)^2$$
$$= 64^2$$
$$= 4096$$

PROBLEM 2-6
Illustrate, using the unit circle model, examples of the following facts:

$$\sin -\theta = -\sin \theta$$
$$\cos -\theta = \cos \theta$$

SOLUTION 2-6
See Fig. 2-4. This shows an example for an angle θ of approximately 60° ($\pi/3$ rad). Note that the angle $-\theta$ is represented by rotation to the same extent as,

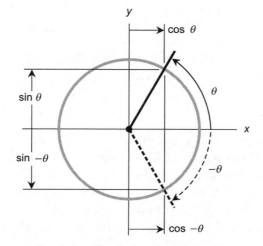

Fig. 2-4. Illustration for Problem 2-6.

but in the opposite direction from, the angle θ. Generally, positive angles are represented by counterclockwise rotation from the x axis, and negative angles are represented by clockwise rotation from the x axis. The ray from the origin for $-\theta$ looks like the reflection of the ray for θ from a pane of glass that contains the x axis and is perpendicular to the page. The above identities can be inferred geometrically from this diagram. The two rays intersect the circle at points whose y values (representing sines) are negatives of each other, and whose x values (representing cosines) are the same.

PROBLEM 2-7
Simplify the expression $\sin(120° - \theta)$. Express coefficients to three decimal places.

SOLUTION 2-7
Use the formula for the sine of an angular difference, given above, substituting 120° for θ in the formula, and θ for ϕ in the formula:

$$\sin(120° - \theta) = (\sin 120°)(\cos \theta) - (\cos 120°)(\sin \theta)$$
$$= 0.866 \cos \theta - (-0.500) \sin \theta$$
$$= 0.866 \cos \theta + 0.500 \sin \theta$$

In case you don't already know this definition, a *coefficient* is a number by which a variable or function is multiplied. In the answer to this problem, the coefficients are 0.866 and 0.500.

PROBLEM 2-8
Illustrate, using the unit circle model, examples of the following facts:

$$\sin(180° - \theta) = \sin \theta$$
$$\cos(180° - \theta) = -\cos \theta$$

SOLUTION 2-8
See Fig. 2-5. This shows an example for an angle θ of approximately 30° ($\pi/6$ rad). The ray from the origin for $180° - \theta$ looks like the reflection of the ray for θ from a pane of glass that contains the y axis and is perpendicular to the page. The above identities can be inferred geometrically from this diagram. The two rays intersect the circle at points whose y values (representing sines) are the same, and whose x values (representing cosines) are negatives of each other.

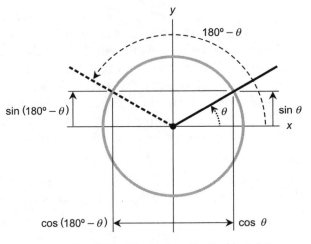

Fig. 2-5.   Illustration for Problem 2-8.

Quiz

Refer to the text in this chapter if necessary. A good score is eight correct. Answers are in the back of the book.

1. Refer to Fig. 2-6. The tangent of $\angle ABC$ is equal to
    (a) the length of line segment AC divided by the length of line segment AB

(b) the length of line segment *AD* divided by the length of line segment *BD*

(c) the length of line segment *AD* divided by the length of line segment *AB*

(d) no ratio of lengths that can be shown here

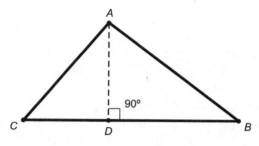

Fig. 2-6.   Illustration for quiz questions.

2. Refer to Fig. 2-6. Suppose we know that the measure of ∠*BCA* is 50° and the length of line segment *AD* is 5.3 units. What is the length of line segment *AC*? Express your answer to one decimal place (that is, the nearest tenth of a unit). Use a calculator if necessary.
   (a) 6.9 units
   (b) 8.2 units
   (c) 6.3 units
   (d) More information is needed to determine the answer

3. Refer again to Fig. 2-6. Suppose we know that the measure of ∠*BCA* is 50° and the length of line segment *AD* is 5.3 units. What is the length of line segment *AB*? Express your answer to one decimal place (that is, the nearest tenth of a unit). Use a calculator if necessary.
   (a) 6.9 units
   (b) 8.2 units
   (c) 6.3 units
   (d) More information is needed to determine the answer

4. Suppose we have a right triangle, and the interior vertex angle at one end of the hypotenuse measures 30°. What is the measure of the interior vertex angle at the other end of the hypotenuse?
   (a) $\pi/3$ rad
   (b) $\pi/4$ rad
   (c) $\pi/6$ rad
   (d) More information is needed to determine the answer

5. Refer again to Fig. 2-6. Suppose we know that line segment AD is exactly 2/3 as long as line segment AB. What is the measure of $\angle DAB$? Express your answer to the nearest tenth of a degree. Use a calculator if necessary.
   (a) 33.7°
   (b) 41.8°
   (c) 48.2°
   (d) 56.3°

6. Suppose, in reference to Fig. 2-6, we are told that the measure of $\angle BCA$ is 50° and the measure of $\angle ABC$ is 38°. We think that the person who says this must be mistaken because
   (a) it would imply that $\triangle ABC$ is a right triangle, which is impossible
   (b) it would imply that the measure of $\angle CAD$ is something other than 40°, but it must be 40° to fulfill the rule that the sum of the measures of the interior angles of any triangle is 180°
   (c) we know that the measure of $\angle ABC$ is 40° because $\triangle ABC$ is an isosceles triangle
   (d) of a rush to judgment! It is entirely possible that the measure of $\angle BCA$ is 50° and the measure of $\angle ABC$ is 38°

7. Suppose you are told that the sine of a certain angle is 0.5299, accurate to four decimal places, and the cosine of that same angle is 0.8480, also accurate to four decimal places. What is the sine of twice this angle, accurate to two decimal places? Don't use the trigonometric function keys on your calculator to figure this out.
   (a) 0.90
   (b) 0.45
   (c) 1.60
   (d) 0.62

8. You are told that the sine of a certain angle is equal to -1.50. You can surmise from this that
   (a) the angle has a measure greater than $\pi/2$ rad but less than π rad
   (b) the angle has a measure greater than π rad but less than $3\pi/2$ rad
   (c) the angle has a measure greater than $3\pi/2$ rad but less than 2π rad
   (d) either you didn't hear the figure correctly, or else the person who told it to you is misinformed

9. Suppose there is a triangle whose sides are 6, 8, and 10 units, respectively. What, approximately, is the tangent of the angle θ opposite the side that measures 8 units?

    (a) 0.600
    (b) 0.750
    (c) 1.333
    (d) It is not defined

10. Suppose there is a triangle whose sides are 6, 8, and 10 units, respectively. What, approximately, is the secant of the angle θ opposite the side that measures 10 units?
    (a) 0.600
    (b) 0.750
    (c) 1.333
    (d) It is not defined

CHAPTER

Graphs and Inverses

Each circular function relates the value of one variable to the value of another, and can be plotted as a graph in rectangular coordinates. Each of the circular functions can be "turned inside-out"; that is, the independent variable and the dependent variable can be interchanged. This gives rise to the *inverse circular functions*. In this chapter, we'll look at the graphs of the circular functions, and also at the graphs of their inverses.

Graphs of Circular Functions

Now that you have begun to get familiar with the use of Greek letters to denote angles, we are going to go back to English letters for a while. In rectangular coordinates, the axes are usually labeled x (for the independent variable) and y (for the dependent variable). Let's use x and y instead of θ and ϕ as the variables when graphing the circular functions. Let's also define the terms *domain of a function* and *range of a function*.

DOMAIN AND RANGE

Suppose f is a function that maps (or assigns) some or all of the elements from a set A to some or all of the elements of a set B. Let A^* be the set of all elements in set A for which there is a corresponding element in set B. Then A^* is called the *domain* of f. Let B^* be the set of all elements in set B for which there is a corresponding element in set A. Then B^* is called the *range* of f.

PROBLEM 3-1

Suppose we take the unit circle, as defined in previous chapters, and cut off its bottom half, but leaving the points $(x,y) = (1,0)$ and $(x,y) = (-1,0)$. This produces a true mathematical function, as opposed to a mere relation, because it ensures that there is never more than one value of y for any value of x. What is the domain of this function?

SOLUTION 3-1

You might want to draw the graph of the unit circle and erase its bottom half, placing a dot at the point $(1,0)$ and another dot at the point $(-1,0)$ to indicate that these points are included in the curve. The domain of this function is represented by the portion of the x axis for which the function is defined. It's easy to see that this is the span of values x such that x is between -1 and 1, inclusive. Formally, if we call A^* the domain of this function, we can write this:

$$A^* = \{x : -1 \leq x \leq 1\}$$

The colon means "such that," and the curly brackets are set notation. So this "mathematese" statement literally reads "A^* equals the set of all real numbers x such that x is greater than or equal to -1 and less than or equal to 1." Sometimes a straight, vertical line is used instead of a colon to mean "such that," so it is also acceptable to write the statement like this:

$$A^* = \{x | -1 \leq x \leq 1\}$$

PROBLEM 3-2

What is the range of the function described above?

SOLUTION 3-2

Look at the drawing you made, showing the graph of the function. The range of this function is represented by the portion of the y axis for which the function is defined: all the values y such that y is between 0 and 1, inclusive. Formally, if we call B^* the range of this function, we can write

$$B^* = \{y : 0 \leq y \leq 1\}$$

GRAPH OF SINE FUNCTION

Figure 3-1 is a graph of the function $y = \sin x$ for values of the domain between $-540°$ and $540°$ (-3π rad and 3π rad). Actually, the domain of the sine function extends over all possible values of x; it is the entire set of real numbers. We limit it here because our page is not infinitely wide! The range of the sine function is limited to values between, and including, -1 and 1. This curve is called a *sine wave* or *sinusoid*. It is significant in electricity, electronics, acoustics, and optics, because it represents an alternating-current (a.c.) signal with all of its energy concentrated at a single frequency.

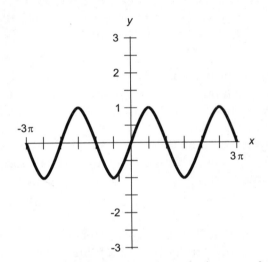

Fig. 3-1. Graph of the sine function for values of x between -3π rad and 3π rad.

GRAPH OF COSINE FUNCTION

Figure 3-2 is a graph of the function $y = \cos x$ for values of the domain between $-540°$ and $540°$ (-3π rad and 3π rad). As is the case with the sine function, the domain of the cosine function extends over the whole set of real numbers. Also like the sine function, the range of the cosine function is limited to values between, and including, -1 and 1. The shape of the *cosine wave* is exactly the same as the shape of the sine wave. Like the sine wave, the cosine wave is *sinusoidal*. The only difference is that the cosine wave is shifted horizontally in the graph by $90°$ ($\pi/2$ rad), or $1/4$ cycle, with respect to the sine wave.

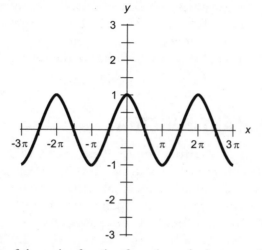

Fig. 3-2.   Graph of the cosine function for values of x between -3π rad and 3π rad.

GRAPH OF TANGENT FUNCTION

Figure 3-3 is a graph of the function $y = \tan x$ for values of the domain between $-540°$ and $540°$ (-3π rad and 3π rad). The range of the tangent function encompasses the entire set of real numbers. But the domain does not! The function "blows up" for certain specific values of x. The "blow-up values" are shown as vertical, dashed lines representing *asymptotes*. For values of x where these asymptotes intersect the x axis, the function $y =$

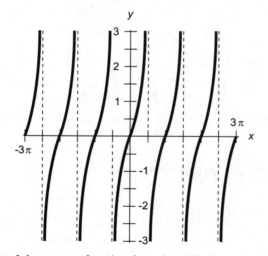

Fig. 3-3.   Graph of the tangent function for values of x between -3π rad and 3π rad.

tan x is undefined. These values, which include all odd integral multiples of 90° ($\pi/2$ rad), are not part of the domain of the tangent function, but all other real numbers are. The term *integral multiple* means that the quantity can be multiplied by any integer, that is, any number in the set {..., −3, −2, −1, 0, 1, 2, 3, ...}.

GRAPH OF COSECANT FUNCTION

Figure 3-4 is a graph of the function $y = \csc x$ for values of the domain between −540° and 540° (−3π rad and 3π rad). The range of the cosecant function encompasses all real numbers greater than or equal to 1, and all real numbers less than or equal to −1. The open interval representing values of y between, but not including, −1 and 1 is not part of the range of this function. The domain includes all real numbers except integral multiples of 180° (π rad). When x is equal to any integral multiple of 180° (π rad), the cosecant function "blows up."

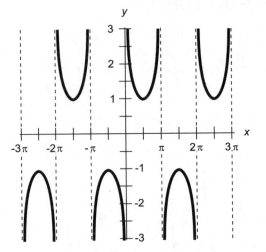

Fig. 3-4.   Graph of the cosecant function for values of x between −3π rad and 3π rad.

GRAPH OF SECANT FUNCTION

Figure 3-5 is a graph of the function $y = \sec x$ for values of the domain between −540° and 540° (−3π rad and 3π rad). The range of the secant function encompasses all real numbers greater than or equal to 1, and all real numbers less than or equal to −1. Thus, the range of the secant function

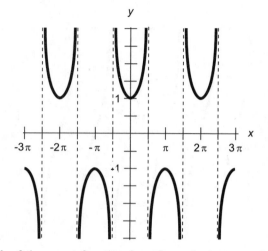

Fig. 3-5.   Graph of the secant function for values of x between -3π rad and 3π rad.

is the same as the range of the cosecant function. But the domain is different. It includes all real numbers except odd integral multiples of 90° ($\pi/2$ rad). The cosecant and secant functions have the same general shape, but they are shifted by 90° ($\pi/2$ rad), or $\frac{1}{4}$ cycle, with respect to each other. This should not come as a surprise, because the cosecant and secant functions are the reciprocals of the sine and cosine functions, respectively, and the sine and cosine are horizontally displaced by $\frac{1}{4}$ cycle.

GRAPH OF COTANGENT FUNCTION

Figure 3-6 is a graph of the function $y = \cot x$ for values of the domain between $-540°$ and $540°$ (-3π rad and 3π rad). The range of the cotangent function encompasses the entire set of real numbers. The domain skips over the integral multiples of 180° (π rad). The graph of the cotangent function looks similar to that of the tangent function. The curves have the same general shape, but while the tangent function always slopes upward as you move toward the right, the cotangent always slopes downward. There is also a *phase shift* of $\frac{1}{4}$ cycle, similar to that which occurs between the cosecant and the secant functions.

PROBLEM 3-3
The domain of the sine function is the same as the domain of the cosine function. In addition, the ranges of the two functions are the same. How can this be true, and yet the two functions are not identical?

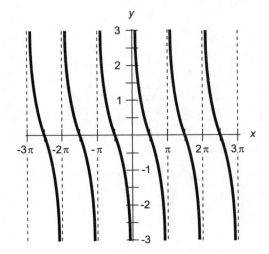

Fig. 3-6.   Graph of the cotangent function for values of x between -3π rad and 3π rad.

SOLUTION 3-3
The difference, as you can see by comparing the graphs of the two functions, is that the curves are displaced along the x axis by $90°$ ($\pi/2$ rad). In general, the cosine of a number is not the same as the sine of that number, although there are certain specific instances in which the two functions have the same value.

PROBLEM 3-4
Draw a graph that shows the specific points where $\sin x = \cos x$.

SOLUTION 3-4
This can be done by superimposing the sine wave and the cosine wave on the same set of coordinates, as shown in Fig. 3-7. The functions attain the same value where the curves intersect.

Inverses of Circular Functions

Each of the circular functions has an inverse: a function that "undoes" whatever the original function does. Defining and working with inverse functions can be tricky.

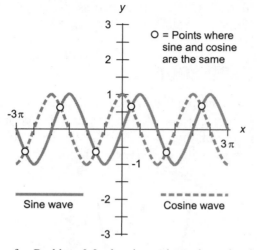

Fig. 3-7.   Illustration for Problem 3-2, showing points where the sine and cosine functions attain the same y value.

WHAT IS AN INVERSE FUNCTION?

What is meant by the term *inverse function*, or *the inverse of a function*? In general terms, the inverse of a function, if it exists, does exactly the reverse of what the function does. We'll get more formal in a moment. But first, we must clarify something about notation.

When a function f has an inverse, it is customary to denote it by a super-script, so it reads f^{-1}. This superscript is not an exponent. The function f^{-1} is not the same thing as the reciprocal of f. If you see $f^{-1}(q)$ written somewhere, it means the inverse function of f applied to the variable q. It does not mean $1/[f(q)]$!

Here is the formal definition. Suppose we have a function f. The inverse of f, call it f^{-1}, is a function such that $f^{-1}[f(x)] = x$ for all x in the domain of f, and $f[f^{-1}(y)] = y$ for all y in the range of f. The function f^{-1} "undoes" what f does, and the function f "undoes" what f^{-1} does. If we apply a function to some value of a variable x and then apply the function's inverse to that, we get x back. If we apply the inverse of a function to some value of a variable y and then apply the original function to that, we get y back.

Not every function has an inverse without some restriction on the domain and/or the range. Sometimes a function f has an inverse f^{-1} without any restrictions; that is, we can simply turn f "inside-out" and get its inverse without worrying about whether this will work for all the values in the domain and range of f. But often, it is necessary to put restrictions on a function in order to be able to define an inverse. Let's look at an example.

SQUARE VS SQUARE ROOT

Figure 3-8 is a graph of a simple function, $f(x) = x^2$. In this graph, the values of $f(x)$ are plotted on the y axis, so we are graphing the equation $y = x^2$. This has a shape familiar to anyone who has taken first-year algebra. It is a parabola opening upward, with its vertex at the origin.

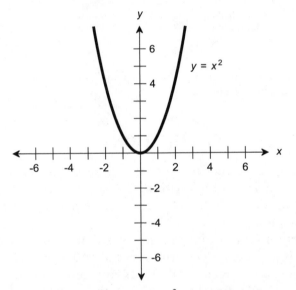

Fig. 3-8.   The relation $y = x^2$ is a function of x.

What do you suppose is the inverse function of f? You might be tempted to say "The square root." If you say that, you're right—partly. Try graphing the parabola with the x and y variables interchanged. You'll plot the curve for the equation $x = y^2$ in that case, and you'll get Fig. 3-9. This is a parabola with exactly the same shape as the one for the equation $y = x^2$, but because the x and y axes are switched, the parabola is turned on its side. This is a perfectly good mathematical relation, and it also happens to be a function that maps values of y to values of x. But it is not a function that maps values of x to values of y. If we call this relation, $g(x) = \pm x^{1/2}$, a function, we are mistaken. We end up with some values of x for which g has no y value (that is okay), and some values of x for which g has two y values (that is not okay). This is easy to see from Fig. 3-9.

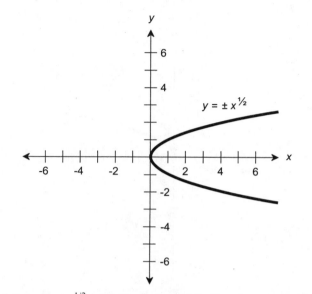

Fig. 3-9.   The relation $y = \pm x^{1/2}$, while the inverse of the function graphed in Fig. 3-8, is not a function.

What can we do to make *g* into a legitimate function? We can require that the *y* values not be negative, and we have a function. Alternatively, we can require that the *y* values not be positive, and again we have a function. Figure 3-10 shows the graph of $y = x^{1/2}$, with the restriction that $y \geq 0$. There exists no abscissa (*x* value) that has more than one ordinate (*y* value).

If you are confused by this, go back to Chapter 1 and review the distinction between a relation and a function.

ARC WHAT?

We can now define the inverses of the circular functions. There are two ways of denoting an inverse when talking about the sine, cosine, tangent, cosecant, secant, and cotangent. We can use the standard abbreviation and add a superscript -1 after it, or we can write "arc" in front of it. Here are the animals, one by one:

- The inverse of the sine function is the arcsine function. If we are operating on some variable *x*, the arcsine of *x* is denoted $\sin^{-1}(x)$ or arcsin (x)

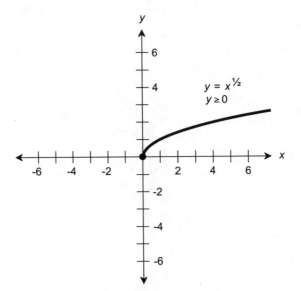

Fig. 3-10.   The relation $y = x^{1/2}$ is a function if we require that y be non-negative.

- The inverse of the cosine function is the arccosine function. If we are operating on some variable x, the arccosine of x is denoted $\cos^{-1}(x)$ or arccos (x)
- The inverse of the tangent function is the arctangent function. If we are operating on some variable x, the arctangent of x is denoted $\tan^{-1}(x)$ or arctan (x)
- The inverse of the cosecant function is the arccosecant function. If we are operating on some variable x, the arccosecant of x is denoted $\csc^{-1}(x)$ or arccsc (x)
- The inverse of the secant function is the arcsecant function. If we are operating on some variable x, the arcsecant of x is denoted $\sec^{-1}(x)$ or arcsec (x)
- The inverse of the cotangent function is the arccotangent function. If we are operating on some variable x, the arccotangent of x is denoted $\cot^{-1}(x)$ or arccot (x)

The sine, cosine, tangent, cosecant, secant, and cotangent require special restrictions in order for the inverses to be definable as legitimate functions. These limits are shown in the graphs of the inverse functions that follow.

USE (AND MISUSE) OF THE −1 SUPERSCRIPT

When using −1 as a superscript in trigonometry, we have to be careful. Ambiguity, or even nonsense, can be the result of improper usage. The expression $\sin^{-1} x$ is not the same thing as $(\sin x)^{-1}$. The former expression refers to the inverse sine of x, or the arcsine of x (arcsin x); but the latter expression means the reciprocal of the sine of x, that is, $1/(\sin x)$. These are not the same. If you have any question about this, plug in a few numbers and test them.

This brings to light an inconsistency in mathematical usage. It is customary to write $(\sin x)^2$ as $\sin^2 x$. But don't try that with the exponent −1, for the reason just demonstrated. You might wonder why the numbers 2 and −1 should be treated so much differently when they are used as superscripts in trigonometry. There is no good answer, except that it is "mathematical convention."

What about other numbers? Does $\sin^{-3} x$, for example, mean the reciprocal of the cube of the sine of x, or the cube of the arcsine of x? Or does it mean the arcsine of the cube of x? If you are worried that the use of a certain notation or expression might produce confusion, don't use it. Use something else, even if it looks less elegant. Saying what you mean is more important than conservation of symbols. It is better to look clumsy and be clear and correct, than to look slick and be ambiguous or mistaken.

PROBLEM 3-5
Is there such a thing as a function that is its own inverse? If so, give one example.

SOLUTION 3-5
The function $f(x) = x$ is its own inverse, and the domain and range both happen to span the entire set of real numbers. If $f(x) = x$, then $f^{-1}(y) = y$. To be sure that this is true, we can check to see if the function "undoes its own action," and that this "undoing operation" works both ways. Let f^{-1} be the inverse of f. We claim that $f^{-1}[f(x)] = x$ for all real numbers x, and $f[f^{-1}(y)] = y$ for all real numbers y. Checking:

$$f^{-1}[f(x)] = f^{-1}(x) = x$$
$$f[f^{-1}(y)] = f(y) = y$$

It works! In fact, it is almost trivial. Why go through such pains to state the obvious? Well, sometimes the obvious turns out to be false, and the wise mathematician or scientist is always wary of this possibility.

PROBLEM 3-6

Find another function that is its own inverse.

SOLUTION 3-6

Consider $g(x) = 1/x$, with the restriction that the domain and range can attain any real-number value except zero. This function is its own inverse; that is, $g^{-1}(x) = 1/x$. To prove this, we must show that $g^{-1}[g(x)] = x$ for all real numbers x except $x = 0$, and also that $g[g^{-1}(y)] = y$ for all real numbers y except $y = 0$. Checking:

$$g^{-1}[g(x)] = g^{-1}(1/x) = 1/(1/x) = x$$
$$g[g^{-1}(y)] = g(1/y) = 1/(1/y) = y$$

It works! This is a little less trivial than the previous example.

PROBLEM 3-7

Find a function for which there exists no inverse function.

SOLUTION 3-7

Consider the function $h(x) = 3$ for all real numbers x. If we try to apply this in reverse, we have to set $y = 3$ in order for $h^{-1}(y)$ to mean anything. Then we end up with all the real numbers at once. Clearly, this is not a function. (Plot a graph of it and see.) Besides this, it is not evident what $h^{-1}(y)$ might be for some value of y other than 3.

Graphs of Circular Inverses

Now that you know what the inverse of a function is, we are ready to look at the graphs of the circular inverses, with the restrictions on the domain and the range necessary to ensure that they are legitimate functions.

GRAPH OF ARCSINE FUNCTION

Figure 3-11 is a graph of the function $y = \arcsin x$ (or $y = \sin^{-1} x$) with its domain limited to values of x between, and including, -1 and 1 (that is, $-1 \le x \le 1$). The range of the arcsine function is limited to values of y between, and including, $-90°$ and $90°$ ($-\pi/2$ rad and $\pi/2$ rad).

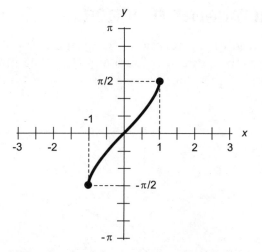

Fig. 3-11.   Graph of the arcsine function for $-1 \leq x \leq 1$.

GRAPH OF ARCCOSINE FUNCTION

Figure 3-12 is a graph of the function $y = \arccos x$ (or $y = \cos^{-1} x$) with its domain limited to values of x between, and including, -1 and 1 (that is, $-1 \leq x \leq 1$). The range of the arccosine function is limited to values of y between, and including, $0°$ and $180°$ (0 rad and π rad).

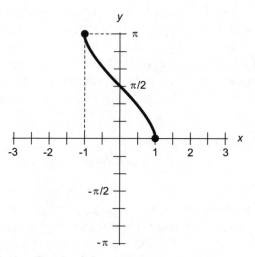

Fig. 3-12.   Graph of the arccosine function for $-1 \leq x \leq 1$.

GRAPH OF ARCTANGENT FUNCTION

Figure 3-13 is a graph of the function $y = \arctan x$ (or $y = \tan^{-1} x$). The domain encompasses the entire set of real numbers. The range of the arctangent function is limited to values of y between, but not including, $-90°$ and $90°$ ($-\pi/2$ and $\pi/2$ rad).

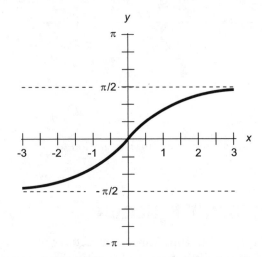

Fig. 3-13.   Graph of the arctangent function for $-3 \leq x \leq 3$.

GRAPH OF ARCCOSECANT FUNCTION

Figure 3-14 is a graph of the function $y = \text{arccsc } x$ (or $y = \csc^{-1} x$) with its domain limited to values of x less than or equal to -1, or greater than or equal to 1 (that is, $x \leq -1$ or $x \geq 1$). The range of the arccosecant function is limited to values of y between, and including, $-90°$ and $90°$ ($-\pi/2$ rad and $\pi/2$ rad), with the exception of $0°$ (0 rad). Mathematically, if R represents the range, we can denote it like this in set notation for degrees and radians, respectively:

$$R = \{y : -90° \leq y < 0° \text{ or } 0° < y \leq 90°\}$$
$$R = \{y : -\pi/2 \leq y < 0 \text{ or } 0 < y \leq \pi/2\}$$

In the latter expression, the "rad" abbreviation is left out. In pure mathematics, the lack of unit specification for angles implies the use of radians by default. If you see angles expressed in mathematical literature and there are no units specified, you should assume that radians are being used, unless the author specifically states otherwise.

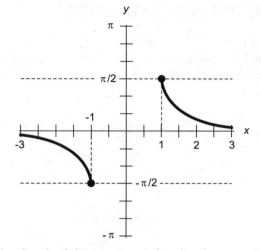

Fig. 3-14. Graph of the arccosecant function for $x \leq -1$ and $x \geq 1$.

GRAPH OF ARCSECANT FUNCTION

Figure 3-15 is a graph of the function $y = \text{arcsec } x$ (or $y = \sec^{-1} x$) with its domain limited to values of x such that $x \leq -1$ or $x \geq 1$. The range of the arcsecant function is limited to values of y such that $0° \leq y < 90°$ or $90° < y \leq 180°$ (0 rad $\leq y < \pi/2$ rad or $\pi/2$ rad $< y \leq \pi$ rad).

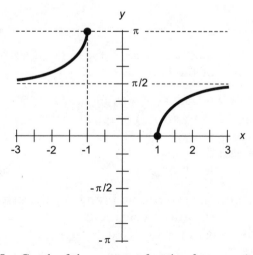

Fig. 3-15. Graph of the arcsecant function for $x \leq -1$ and $x \geq 1$.

GRAPH OF ARCCOTANGENT FUNCTION

Figure 3-16 is a graph of the function $y = \text{arccot } x$ (or $y = \cot^{-1} x$). Its domain encompasses the entire set of real numbers. The range of the arccotangent function is limited to values of y between, but not including, $0°$ and $180°$ (0 rad and π rad).

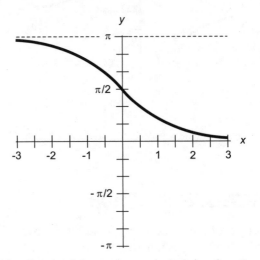

Fig. 3-16.   Graph of the arccotangent function for $-3 \le x \le 3$.

Quiz

Refer to the text in this chapter if necessary. A good score is eight correct. Answers are in the back of the book.

1. The sine function and the tangent function
    (a) have identical shapes when graphed
    (b) have different ranges
    (c) have identical domains
    (d) are inverses of each other

2. The restrictions on the domain and range of the inverse circular functions are necessary in order to ensure that:
    (a) no negative angles are involved
    (b) they never "blow up"

(c) none of them have more than one y value (ordinate) for any x value (abscissa)

(d) the domains are defined for all real numbers

3. The graph of the cosine function
    (a) has the same shape as the graph of the sine function, but is "stretched" vertically
    (b) has the same shape as the graph of the sine function, but is shifted horizontally
    (c) has the same shape as the graph of the sine function, but is shifted vertically
    (d) has the same shape as the graph of the sine function, but is "squashed" horizontally

4. The domain of the arccotangent function
    (a) encompasses only the real numbers between, and including, -1 and 1
    (b) encompasses only the values between $90°$ ($\pi/2$ rad) and $270°$ ($3\pi/2$ rad)
    (c) encompasses only the real numbers less than -1 or greater than 1
    (d) encompasses all of the real numbers

5. What does the expression $(\sin x)^{-1}$ denote?
    (a) The reciprocal of the sine of x
    (b) The sine of $1/x$
    (c) The arcsine of x
    (d) None of the above

6. Look at Fig. 3-11. Consider the interval S of all values of y such that y is between, and including, -1 rad and 1 rad. Which of the following statements is true?
    (a) S constitutes part of the domain of the function shown in the graph
    (b) S constitutes part of the range of the function shown in the graph
    (c) S constitutes all of the domain of the function shown in the graph
    (d) S constitutes all of the range of the function shown in the graph

7. Look at Fig. 3-13. What can be said about this function based on its appearance in the graph?
    (a) Its range is limited
    (b) Its range spans the entire set of real numbers
    (c) Its domain is limited
    (d) It is not, in fact, a legitimate function

8. The graph of the function $y = \sin x$ "blows up" at
   (a) all values of x that are multiples of $90°$
   (b) all values of x that are odd multiples of $90°$
   (c) all values of x that are even multiples of $90°$
   (d) no values of x

9. The function $y = \csc x$ is defined for
   (a) only those values of x less than -1 or greater than 1
   (b) only those values of x between, and including, -1 and 1
   (c) all values of x except integral multiples of π rad
   (d) all values of x except integral multiples of $\pi/2$ rad

10. Which of the following graphs does not "blow up" for any value of x?
    (a) The curve for $y = \arctan x$
    (b) The curve for $y = \tan x$
    (c) The curve for $y = \text{arccsc } x$
    (d) The curve for $y = \csc x$

4

Hyperbolic Functions

There are six *hyperbolic functions* that are similar in some ways to the circular functions. They are known as the *hyperbolic sine*, *hyperbolic cosine*, *hyperbolic tangent*, *hyperbolic cosecant*, *hyperbolic secant*, and *hyperbolic cotangent*. In formulas and equations, they are abbreviated sinh, cosh, tanh, csch, sech, and coth respectively.

The hyperbolic functions are based on certain characteristics of the *unit hyperbola*, which has the equation $x^2 - y^2 = 1$ in rectangular coordinates. Hyperbolic functions are used in certain engineering applications.

You can have fun trying to pronounce the abbreviations for hyperbolic functions (but not with your mouth full of food); but it is best to name a hyperbolic function straightaway when talking about it. For example, when you see "tanh," say "hyperbolic tangent."

The Hyper Six

The circular functions operate on angles. In theory, the hyperbolic functions do too. Units are generally not mentioned for the quantities on which the hyperbolic functions operate, but they are understood to be in radians. Greek symbols are not always used to denote these variables. Plain lowercase

English italicized x and y are common. Some mathematicians, scientists, and engineers prefer to use u and v. Once in a while you'll come across a paper where the author uses the lowercase italicized Greek alpha (α) and beta (β) to represent the angles in hyperbolic functions.

POWERS OF e

Once we define the hyperbolic sine and the hyperbolic cosine of a quantity, the other four hyperbolic functions can be defined, just as the circular tangent, cosecant, secant, and cotangent follow from the circular sine and cosine.

In order to clearly define what is meant by the hyperbolic sine and the hyperbolic cosine, we use *base-e exponential functions*. These revolve around a number that is denoted e. This number has some special properties. It is an *irrational number*—a number that can't be precisely expressed as a ratio of two whole numbers. The best we can do is approximate it. (The term "irrational," in mathematics, means "not expressible as a ratio of whole numbers." It does not mean "unreasonable" or "crazy.")

If you have a calculator with a function key marked "e^x" you can determine the value of e to several decimal places by entering the number 1 and then hitting the "e^x" key. If your calculator does not have an "e^x" key, it should have a key marked "ln" which stands for *natural logarithm*, and a key marked "inv" which stands for *inverse*. To get e from these keys, enter the number 1, and then hit "inv" and "ln" in succession. You should get a number whose first few digits are 2.71828.

If you want to determine the value of e^x for some quantity x other than 1, you should enter the value x and then hit either the "e^x" key or else hit the "inv" and "ln" keys in succession, depending on the type of calculator you have. In order to find e^{-x}, find e^x first, and then find the reciprocal of this by hitting the "$1/x$" key.

If your calculator lacks exponential or natural logarithm functions, it is time for you to go out and buy one. Most personal computers have calculator programs that can be placed in "scientific mode," where these functions are available.

TWO TO START

Let x be a real number. The hyperbolic sine and the hyperbolic cosine can be defined in terms of powers of e, like this:

$$\sinh x = (e^x - e^{-x})/2$$
$$\cosh x = (e^x + e^{-x})/2$$

If these look intimidating, just remember that using them involves nothing more than entering numbers into a calculator and hitting certain keys in the correct sequence.

In a theoretical course, you will find other ways of expressing the hyperbolic sine and cosine functions, but for our purposes, the above two formulas are sufficient.

THE OTHER FOUR

The remaining four hyperbolic functions follow from the hyperbolic sine and the hyperbolic cosine, like this:

$$\tanh x = \sinh x / \cosh x$$
$$\operatorname{csch} x = 1/\sinh x$$
$$\operatorname{sech} x = 1/\cosh x$$
$$\coth x = \cosh x / \sinh x$$

In terms of exponential functions, they are expressed this way:

$$\tanh x = (e^x - e^{-x})/(e^x + e^{-x})$$
$$\operatorname{csch} x = 2/(e^x - e^{-x})$$
$$\operatorname{sech} x = 2/(e^x + e^{-x})$$
$$\coth x = (e^x + e^{-x})/(e^x - e^{-x})$$

Now let's look at the graphs of the six hyperbolic functions. As is the case with the inverses of the circular functions, the domain and/or range of the inverse of a hyperbolic function may have to be restricted to ensure that there is never more than one ordinate (y value) for a given abscissa (x value).

HYPERBOLIC SINE

Figure 4-1 is a graph of the function $y = \sinh x$. Its domain and range both extend over the entire set of real numbers.

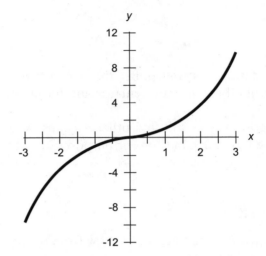

Fig. 4-1.   Graph of the hyperbolic sine function.

HYPERBOLIC COSINE

Figure 4-2 is a graph of the function $y = \cosh x$. Its domain extends over the whole set of real numbers, and its range is the set of real numbers y greater than or equal to 1.

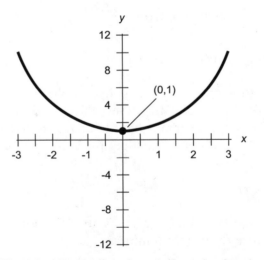

Fig. 4-2.   Graph of the hyperbolic cosine function.

HYPERBOLIC TANGENT

Figure 4-3 is a graph of the function $y = \tanh x$. Its domain encompasses the entire set of real numbers. The range of the hyperbolic tangent function is limited to the set of real numbers y between, but not including, -1 and 1; that is, $-1 < y < 1$.

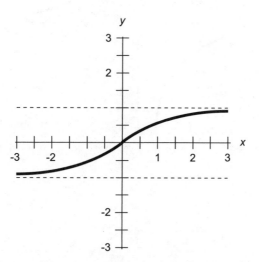

Fig. 4-3.   Graph of the hyperbolic tangent function.

HYPERBOLIC COSECANT

Figure 4-4 is a graph of the function $y = \operatorname{csch} x$. Its domain encompasses the set of real numbers x such that $x \neq 0$. The range of the hyperbolic cotangent function encompasses the set of real numbers y such that $y \neq 0$.

HYPERBOLIC SECANT

Figure 4-5 is a graph of the function $y = \operatorname{sech} x$. Its domain encompasses the entire set of real numbers. Its range is limited to the set of real numbers y greater than 0 but less than or equal to 1; that is, $0 < y \leq 1$.

HYPERBOLIC COTANGENT

Figure 4-6 is an approximate graph of the function $y = \coth x$. Its domain encompasses the entire set of real numbers x such that $x \neq 0$. The range of

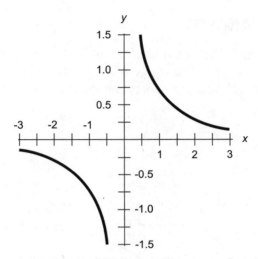

Fig. 4-4.   Graph of the hyperbolic cosecant function.

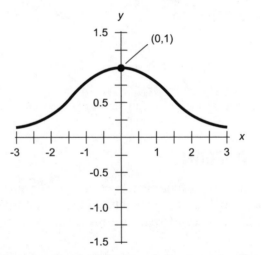

Fig. 4-5.   Graph of the hyperbolic secant function.

the hyperbolic cotangent function encompasses the set of real numbers y less than -1 or greater than 1; that is, $y < -1$ or $y > 1$.

PROBLEM 4-1
Why does the graph of $y = \operatorname{csch} x$ "blow up" when $x = 0$? Why is $\operatorname{csch} x$ not defined when $x = 0$?

SOLUTION 4-1
Remember that the hyperbolic cosecant (csch) is the reciprocal of the hyperbolic sine (sinh). If $x = 0$, then $\sinh x = 0$, as you can see from Fig. 4-1. As x

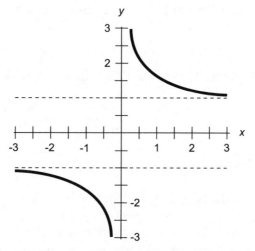

Fig. 4-6.   Graph of the hyperbolic cotangent function.

approaches zero (written $x \to 0$) from either side, the value of the hyperbolic sine also approaches zero (sinh $x \to 0$). Thus, csch x, which is equal to $1/(\sinh x)$ and is graphed in Fig. 4-4, grows without limit as $x \to 0$ from either direction. The value of y "blows up" positively as $x \to 0$ from the positive, or right, side (written $x \to 0^+$) and negatively as $x \to 0$ from the negative, or left, side ($x \to 0^-$). When $x = 0$, the reciprocal of the hyperbolic sine is not defined, because it is a quotient with 0 in the denominator.

PROBLEM 4-2
What is the hyperbolic cotangent of 0? Express it in two ways.

SOLUTION 4-2
This quantity is not defined. The easiest way to demonstrate this fact is to look at the graph of the hyperbolic cotangent function (Fig. 4-6). The graph of the function $y = \coth x$ "blows up" at $x = 0$. It doesn't have a y value there.

We can also express coth 0 by first finding the values of sinh 0 and cosh 0 using the exponential definitions. Remember the formulas:

$$\sinh x = (e^x - e^{-x})/2$$
$$\cosh x = (e^x + e^{-x})/2$$

If $x = 0$, then $e^x = 1$ and $e^{-x} = 1$. Therefore:

$$\sinh 0 = (1 - 1)/2 = 0/2 = 0$$
$$\cosh 0 = (1 + 1)/2 = 2/2 = 1$$

The hyperbolic cotangent is the hyperbolic cosine divided by the hyperbolic sine:

$$\coth 0 = \cosh 0 / \sinh 0 = 1/0$$

This expression is undefined, because it is a quotient with 0 in the denominator.

Hyperbolic Inverses

Each of the six hyperbolic functions has an inverse relation. These are known as the *hyperbolic arcsine, hyperbolic arccosine, hyperbolic arctangent, hyperbolic arccosecant, hyperbolic arcsecant,* and *hyperbolic arccotangent.* In formulas and equations, they are abbreviated arcsinh or $\sinh^{-1}$, arccosh or $\cosh^{-1}$, arctanh or $\tanh^{-1}$, arccsch or csch^{-1}, arcsech or sech^{-1}, and arccoth or $\coth^{-1}$ respectively. These relations become functions when their domains are restricted as shown in the graphs of Figs. 4-7 through 4-12.

THE NATURAL LOGARITHM

Now it is time to learn a little about logarithms. It is common to write "the natural logarithm of x" as "ln x." This function is the inverse of the base-e exponential function. The natural logarithm function and the base-e exponential function "undo" each other. Suppose x and v are real numbers, and y and u are positive real numbers. If $e^x = y$, then $x = \ln y$, and if $\ln u = v$, then $u = e^v$.

The natural logarithm function is useful in expressing the inverse hyperbolic functions, just as the exponential function can be used to express the hyperbolic functions.

You can find the natural logarithm of a specific number using a calculator. Enter the number for which you want to find the natural logarithm, and then hit the "ln" key. Beware: the logarithm of 0 or any negative real number is not defined in the set of real numbers.

HYPERBOLIC INVERSES AS LOGARITHMS

You can find hyperbolic inverses of specific quantities using a calculator that has the "ln" function. Here are the expressions for the hyperbolic inverses, in terms of natural logarithms. (The $1/2$ power represents the square root.)

$$\text{arcsinh } x = \ln \left[x + (x^2 + 1)^{1/2}\right]$$
$$\text{arccosh } x = \ln \left[x + (x^2 - 1)^{1/2}\right]$$
$$\text{arctanh } x = 0.5 \ln \left[(1 + x)/(1 - x)\right]$$
$$\text{arccsch } x = \ln \left[x^{-1} + (x^{-2} + 1)^{1/2}\right]$$
$$\text{arcsech } x = \ln \left[x^{-1} + (x^{-2} - 1)^{1/2}\right]$$
$$\text{arccoth } x = 0.5 \ln \left[(x + 1)/(x - 1)\right]$$

In these expressions, the values 0.5 represent exactly $1/2$. The formulas are a little bit messy, but if you plug in the numbers and take your time doing the calculations, you shouldn't have trouble. Be careful about the order in which you perform the operations. Perform the operations in the innermost sets of parentheses or brackets first, and then work outward.

Let's see what the graphs of the inverse hyperbolic functions look like.

HYPERBOLIC ARCSINE

Figure 4-7 is a graph of the function $y = \text{arcsinh } x$ (or $y = \sinh^{-1} x$). Its domain and range both encompass the entire set of real numbers.

HYPERBOLIC ARCCOSINE

Figure 4-8 is a graph of the function $y = \text{arccosh } x$ (or $y = \cosh^{-1} x$). The domain includes real numbers x such that $x \geq 1$. The range of the hyperbolic

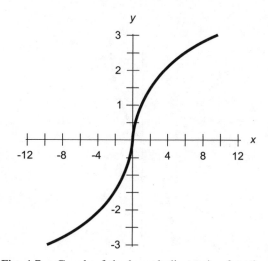

Fig. 4-7. Graph of the hyperbolic arcsine function.

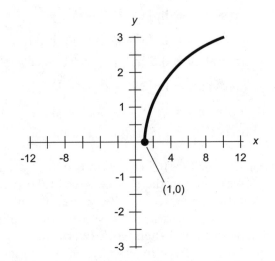

Fig. 4-8.   Graph of the hyperbolic arccosine function.

arccosine function is limited to the non-negative reals, that is, to real numbers y such that $y \geq 0$.

HYPERBOLIC ARCTANGENT

Figure 4-9 is a graph of the function $y = \text{arctanh } x$ (or $y = \tanh^{-1} x$). The domain is limited to real numbers x such that $-1 < x < 1$. The range of the hyperbolic arctangent function spans the entire set of real numbers.

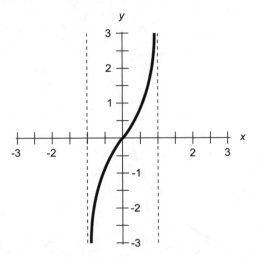

Fig. 4-9.   Graph of the hyperbolic arctangent function.

HYPERBOLIC ARCCOSECANT

Figure 4-10 is a graph of the function $y = \text{arccsch } x$ (or $y = \text{csch}^{-1} x$). Both the domain and the range of the hyperbolic arccosecant function include all real numbers except zero.

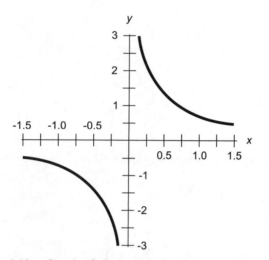

Fig. 4-10.   Graph of the hyperbolic arccosecant function.

HYPERBOLIC ARCSECANT

Figure 4-11 is a graph of the function $y = \text{arcsech } x$ (or $y = \text{sech}^{-1} x$). The domain of this function is limited to real numbers x such that $0 < x \leq 1$. The range of the hyperbolic arcsecant function is limited to the non-negative reals, that is, to real numbers y such that $y \geq 0$.

HYPERBOLIC ARCCOTANGENT

Figure 4-12 is a graph of the function $y = \text{arccoth } x$ (or $y = \text{coth}^{-1} x$). The domain of this function includes all real numbers x such that $x < -1$ or $x > 1$. The range of the hyperbolic arccotangent function includes all real numbers except zero.

PROBLEM 4-3
What is the value of arcsinh 0? Use a calculator if you need it.

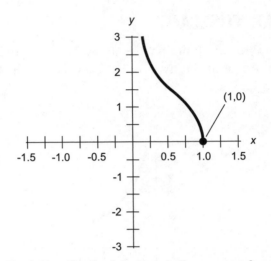

Fig. 4-11.   Graph of the hyperbolic arcsecant function.

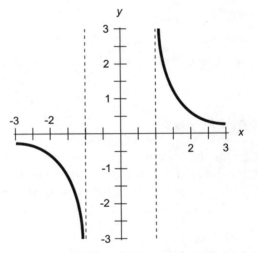

Fig. 4-12.   Graph of the hyperbolic arccotangent function.

SOLUTION 4-3

From the graph in Fig. 4-7, it appears that it ought to be 0. We can verify this by using the formula above along with a calculator if needed:

$$\text{arcsinh } x = \ln \left[x + (x^2 + 1)^{1/2} \right]$$

$$\text{arcsinh } 0 = \ln \left[0 + (0^2 + 1)^{1/2} \right]$$

$$= \ln (0 + 1^{1/2})$$
$$= \ln (0 + 1)$$
$$= \ln 1$$
$$= 0$$

If you've had any experience with logarithms, you don't need a calculator to do the above calculation, because you already know that the natural logarithm of 1 is equal to 0.

PROBLEM 4-4
What is the value of arccsch 1? Use a calculator if you need it. Use the logarithm-based formulas to determine the answer, and express it to three decimal places.

SOLUTION 4-4
From the graph in Fig. 4-10, we can guess that arccsch 1 ought to be a little less than 1. Let's use the formula above and find out:

$$\text{arccsc } h \; x = \ln [x^{-1} + (x^{-2} + 1)^{1/2}]$$
$$\text{arccsc } h \; 1 = \ln [1^{-1} + (1^{-2} + 1)^{1/2}]$$
$$= \ln [1 + (1 + 1)^{1/2}]$$
$$= \ln (1 + 2^{1/2})$$
$$= \ln (1 + 1.41421)$$
$$= \ln 2.41421$$
$$= 0.881 \text{ (rounded to three decimal places)}$$

Hyper Facts

Here's another flurry of facts, this time involving the hyperbolic functions. You are not expected to memorize any of these, but you should be able to use them in calculations if you are given numbers to "plug in."

PYTHAGOREAN THEOREM FOR SINH AND COSH

The difference between the squares of the hyperbolic sine and hyperbolic cosine of a variable is always equal to either 1 or -1. The following formulas hold for all real numbers x:

$$\sinh^2 x - \cosh^2 x = -1$$
$$\cosh^2 x - \sinh^2 x = 1$$

PYTHAGOREAN THEOREM FOR CSCH AND COTH

The difference between the squares of the hyperbolic cotangent and hyperbolic cosecant of a variable is always equal to either 1 or −1. The following formulas hold for all real numbers x except 0:

$$\operatorname{csch}^2 x - \coth^2 x = -1$$
$$\coth^2 x - \operatorname{csch}^2 x = 1$$

PYTHAGOREAN THEOREM FOR SECH AND TANH

The sum of the squares of the hyperbolic secant and hyperbolic tangent of a variable is always equal to 1. The following formula holds for all real numbers x:

$$\operatorname{sech}^2 x + \tanh^2 x = 1$$

HYPERBOLIC SINE OF NEGATIVE VARIABLE

The hyperbolic sine of the negative of a variable is equal to the negative of the hyperbolic sine of the variable. The following formula holds for all real numbers x:

$$\sinh \, -x = -\sinh x$$

HYPERBOLIC COSINE OF NEGATIVE VARIABLE

The hyperbolic cosine of the negative of a variable is equal to the hyperbolic cosine of the variable. The following formula holds for all real numbers x:

$$\cosh \, -x = \cosh x$$

HYPERBOLIC TANGENT OF NEGATIVE VARIABLE

The hyperbolic tangent of the negative of a variable is equal to the negative of the hyperbolic tangent of the variable. The following formula holds for all real numbers x:

$$\tanh -x = -\tanh x$$

HYPERBOLIC COSECANT OF NEGATIVE VARIABLE

The hyperbolic cosecant of the negative of a variable is equal to the negative of the hyperbolic cosecant of the variable. The following formula holds for all real numbers x except 0:

$$\operatorname{csch} -x = -\operatorname{csch} x$$

HYPERBOLIC SECANT OF NEGATIVE VARIABLE

The hyperbolic secant of the negative of a variable is equal to the hyperbolic secant of the variable. The following formula holds for all real numbers x:

$$\operatorname{sech} -x = \operatorname{sech} x$$

HYPERBOLIC COTANGENT OF NEGATIVE VARIABLE

The hyperbolic cotangent of the negative of a variable is equal to the negative of the hyperbolic cotangent of the variable. The following formula holds for all real numbers x except 0:

$$\coth -x = -\coth x$$

HYPERBOLIC SINE OF DOUBLE VALUE

The hyperbolic sine of twice any given variable is equal to twice the hyperbolic sine of the original variable times the hyperbolic cosine of the original variable. The following formula holds for all real numbers x:

$$\sinh 2x = 2 \sinh x \cosh x$$

HYPERBOLIC COSINE OF DOUBLE VALUE

The hyperbolic cosine of twice any given variable can be found according to any of the following three formulas for all real numbers x:

$$\cosh 2x = \cosh^2 x + \sinh^2 x$$
$$\cosh 2x = 1 + 2 \sinh^2 x$$
$$\cosh 2x = 2 \cosh^2 x - 1$$

HYPERBOLIC TANGENT OF DOUBLE VALUE

The hyperbolic tangent of twice a given variable can be found according to the following formula for all real numbers x:

$$\tanh 2x = (2 \tanh x)/(1 + \tanh^2 x)$$

HYPERBOLIC SINE OF HALF VALUE

The hyperbolic sine of half any given variable can be found according to the following formula for all non-negative real numbers x:

$$\sinh (x/2) = [(1 - \cosh x)/2]^{1/2}$$

For negative real numbers x, the formula is:

$$\sinh (x/2) = -[(1 - \cosh x)/2]^{1/2}$$

HYPERBOLIC COSINE OF HALF VALUE

The hyperbolic cosine of half any given variable can be found according to the following formula for all real numbers x:

$$\cosh (x/2) = [(1 + \cosh x)/2]^{1/2}$$

HYPERBOLIC SINE OF SUM

The hyperbolic sine of the sum of two variables x and y can be found according to the following formula for all real numbers x and y:

$$\sinh (x + y) = \sinh x \cosh y + \cosh x \sinh y$$

HYPERBOLIC COSINE OF SUM

The hyperbolic cosine of the sum of two variables x and y can be found according to the following formula for all real numbers x and y:

$$\cosh (x + y) = \cosh x \cosh y + \sinh x \sinh y$$

HYPERBOLIC TANGENT OF SUM

The hyperbolic tangent of the sum of two variables x and y can be found according to the following formula for all real numbers x and y:

$$\tanh (x + y) = (\tanh x + \tanh y)/(1 + \tanh x \tanh y)$$

HYPERBOLIC SINE OF DIFFERENCE

The hyperbolic sine of the difference between two variables x and y can be found according to the following formula for all real numbers x and y:

$$\sinh (x - y) = \sinh x \cosh y - \cosh x \sinh y$$

HYPERBOLIC COSINE OF DIFFERENCE

The hyperbolic cosine of the difference between two variables x and y can be found according to the following formula for all real numbers x and y:

$$\cosh (x - y) = \cosh x \cosh y - \sinh x \sinh y$$

HYPERBOLIC TANGENT OF DIFFERENCE

The hyperbolic tangent of the difference between two variables x and y can be found according to the following formula for all real numbers x and y, provided the product of $\tanh x$ and $\tanh y$ is not equal to 1:

$$\tanh (x - y) = (\tanh x - \tanh y)/(1 - \tanh x \tanh y)$$

PROBLEM 4-5
Based on the above formulas, find a formula for the hyperbolic sine of three times a given value. That is, find a general formula for $\sinh 3x$. Express the answer in terms of functions of x only.

SOLUTION 4-5

Let's start out by supposing that $y = 2x$, so $x + y = x + 2x = 3x$. We have a formula for the hyperbolic sine of the sum of two values. It is:

$$\sinh (x + y) = \sinh x \cosh y + \cosh x \sinh y$$

Substituting $2x$ in place of y, we know this:

$$\sinh 3x = \sinh (x + 2x) = \sinh x \cosh 2x + \cosh x \sinh 2x$$

We have formulas to determine $\cosh 2x$ and $\sinh 2x$. They are:

$$\cosh 2x = \cosh^2 x + \sinh^2 x$$
$$\sinh 2x = 2 \sinh x \cosh x$$

We can substitute these equivalents in the previous formula, getting this:

$$\sinh 3x = \sinh x (\cosh^2 x + \sinh^2 x) + \cosh x (2 \sinh x \cosh x)$$
$$= \sinh x \cosh^2 x + \sinh^3 x + 2 \sinh x \cosh^2 x$$
$$= 3 \sinh x \cosh^2 x + \sinh^3 x$$

There are two other ways this problem can be solved, because there are three different formulas for the hyperbolic cosine of a double value.

PROBLEM 4-6

Verify (approximately) the following formula for $x = 3$ and $y = 2$:

$$\sinh (x - y) = \sinh x \cosh y - \cosh x \sinh y$$

SOLUTION 4-6

Let's plug in the numbers:

$$\sinh (3 - 2) = \sinh 3 \cosh 2 - \cosh 3 \sinh 2$$
$$\sinh 1 = \sinh 3 \cosh 2 - \cosh 3 \sinh 2$$

Using a calculator, we find these values based on the exponential formulas for the hyperbolic sine and cosine:

$$\sinh 1 = 1.1752$$
$$\sinh 2 = 3.6269$$
$$\sinh 3 = 10.0179$$
$$\cosh 2 = 3.7622$$
$$\cosh 3 = 10.0677$$

We can put these values into the second formula above and see if the numbers add up. We should find that the following expression calculates out to approximately sinh 1, or 1.1752. Here we go:

$$10.0179 \times 3.7622 - 10.0677 \times 3.6269$$
$$= 37.689 - 36.515$$
$$= 1.174$$

This is close enough, considering that *error accumulation* occurs when performing repeated calculations with numbers that aren't exact. Error accumulation involves the idiosyncrasies of *scientific notation* and *significant figures*. When significant figures aren't taken seriously, they (or their lack) can cause trouble for experimental scientists, engineers, surveyors, and navigators. You'll learn about scientific notation and significant figures in Chapter 7.

Quiz

Refer to the text in this chapter if necessary. A good score is eight correct. Answers are in the back of the book.

1. Suppose we know that the hyperbolic cosine of a certain variable is equal to 1. What is the hyperbolic cosine of twice that variable?
   (a) 1
   (b) 0
   (c) e^{-1}
   (d) e

2. From the logarithm formulas, it is apparent that the hyperbolic arctangent of 1 is
   (a) equal to e
   (b) equal to $1/e$
   (c) equal to 0
   (d) not defined

3. The number e is equal to the ratio of
   (a) a circle's area to its radius
   (b) a circle's diameter to its radius
   (c) two large negative integers
   (d) no two whole numbers

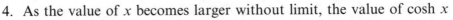
4. As the value of x becomes larger without limit, the value of cosh x
   (a) also becomes larger without limit
   (b) approaches zero
   (c) approaches 1
   (d) becomes larger without limit, negatively

5. How is the hyperbolic secant of 10 related to the hyperbolic secant of -10?
   (a) They are reciprocals
   (b) They add up to zero
   (c) Their ratio is equal to e
   (d) They are the same

6. A unit hyperbola can be represented by the equation
   (a) $x^2 = 1 + y^2$
   (b) $x^2 + y^2 = 1$
   (c) $y = 1 - x^2$
   (d) $y = 1 + x^2$

7. A simpler way to express e to the power of arcsech x is
   (a) non-existent because such an expression is too complicated to deal with
   (b) $x^{-1} + (x^{-2} - 1)^{1/2}$
   (c) $x^{-1} + (x^{-2} + 1)^{1/2}$
   (d) $x^{-1} - (x^{-2} - 1)^{1/2}$

8. As the value of x increases without limit, what happens to the value of e^{-x}?
   (a) It becomes larger and larger, positively
   (b) It stays the same
   (c) It approaches zero
   (d) It becomes larger and larger, negatively

9. The hyperbolic tangent is equivalent to
   (a) the reciprocal of the hyperbolic sine
   (b) the ratio of the hyperbolic sine to the hyperbolic cosine
   (c) the reciprocal of the circular sine
   (d) the ratio of the circular sine to the circular cosine

10. Using the exponential formulas, the hyperbolic sine of 3 is expressed as
   (a) $(e^3 - e^{-3})/2$
   (b) $2/(e^3 - e^{-3})$
   (c) $(e^3 + e^{-3})/2$
   (d) any of the above

Polar Coordinates

The Cartesian scheme is not the only way that points can be located on a flat surface. Instead of moving right–left and up–down from an origin point, we can travel outward a certain distance, and in a certain direction, from that point. The outward distance is called the *radius* or *range*. It is measured in linear units, either arbitrary or specific (such as meters or kilometers). The direction is measured in angular units (either radians or degrees). It is sometimes called the *azimuth*, *bearing*, or *heading*.

The Mathematician's Way

The *polar coordinate plane*, as used by mathematicians and also by some engineers, is shown in Figs. 5-1 and 5-2. The independent variable is plotted as an angle θ relative to a reference axis pointing to the right (or "east"), and the dependent variable is plotted as the distance or radius r from the origin. Coordinate points are thus denoted as ordered pairs (θ, r).

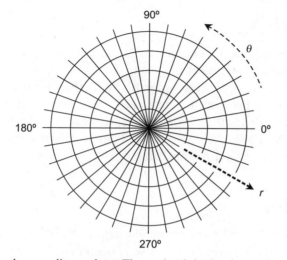

Fig. 5-1.   The polar coordinate plane. The angle θ is in degrees, and the radius r is in arbitrary units.

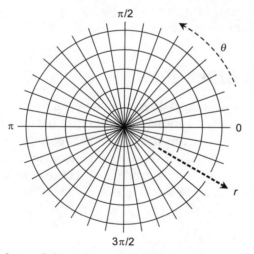

Fig. 5-2.   Another form of the polar coordinate plane. The angle θ is in radians, and the radius r is in arbitrary units.

THE RADIUS

In any polar plane, the radii are shown by concentric circles. The larger the circle, the greater the value of r. In Figs. 5-1 and 5-2, the circles are not labeled in units. Imagine each concentric circle, working outward, as increasing by any number of units you want. For example, each radial division might represent one unit, or five units, or 10, or 100.

THE DIRECTION

Direction can be expressed in degrees or radians counterclockwise from a reference axis pointing to the right or "east." In Fig. 5-1, the direction θ is in degrees. Figure 5-2 shows the same polar plane, using radians to express the direction. (The "rad" abbreviation is not used, because it is obvious from the fact that the angles are multiples of π.) Regardless of whether degrees or radians are used, the angular scale is linear. That is, the physical angle on the graph is directly proportional to the value of θ.

NEGATIVE RADII

In polar coordinates, it is all right to have a negative radius. If some point is specified with $r < 0$, we multiply r by -1 so it becomes positive, and then add or subtract $180°$ (π rad) to or from the direction. That's like saying, "Proceed 10 kilometers east" instead of "Proceed negative 10 kilometers west." Negative radii are allowed in order to graph figures that represent functions whose ranges can attain negative values.

NON-STANDARD DIRECTIONS

It's all right to have non-standard direction angles in polar coordinates. If the value of θ is $360°$ (2π rad) or more, it represents more than one complete counterclockwise revolution from the $0°$ (0 rad) reference axis. If the direction angle is less than $0°$ (0 rad), it represents clockwise revolution instead of counterclockwise revolution. Non-standard direction angles are allowed in order to graph figures that represent functions whose domains go outside the standard angle range.

Some Examples

To see how the polar coordinate system works, let's look at the graphs of some familiar objects. Circles, ellipses, spirals, and other figures whose equations are complicated in Cartesian coordinates can often be expressed much more simply in polar coordinates. In general, the polar direction θ is expressed in radians. In the examples that follow, the "rad" abbreviation is eliminated, because it is understood that all angles are in radians.

CIRCLE CENTERED AT ORIGIN

The equation of a circle centered at the origin in the polar plane is given by the following formula:

$$r = a$$

where a is a real-number constant greater than 0. This is illustrated in Fig. 5-3.

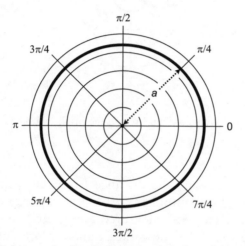

Fig. 5-3.   Polar graph of a circle centered at the origin, with radius a.

CIRCLE PASSING THROUGH ORIGIN

The general form for the equation of a circle passing through the origin and centered at the point (θ_0, r_0) in the polar plane (Fig. 5-4) is as follows:

$$r = 2r_0 \cos (\theta - \theta_0)$$

ELLIPSE CENTERED AT ORIGIN

The equation of an ellipse centered at the origin in the polar plane is given by the following formula:

$$r = ab/(a^2 \sin^2 \theta + b^2 \cos^2 \theta)^{1/2}$$

where a and b are real-number constants greater than 0.

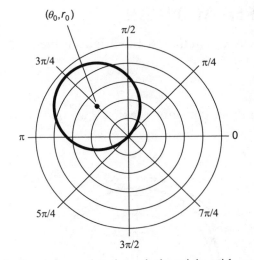

Fig. 5-4.   Polar graph of a circle passing through the origin, with center at (θ_0, r_0) and radius r_0.

In the ellipse, a represents the distance from the origin to the curve as measured along the "horizontal" ray $\theta = 0$, and b represents the distance from the origin to the curve as measured along the "vertical" ray $\theta = \pi/2$. This is illustrated in Fig. 5-5. The values a and b represent the lengths of the *semi-axes* of the ellipse. The greater value is the length of the *major semi-axis*, and the lesser value is the length of the *minor semi-axis*.

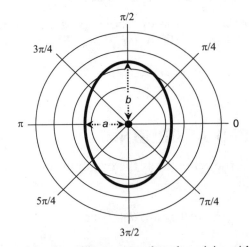

Fig. 5-5.   Polar graph of an ellipse centered at the origin, with semi-axes a and b.

HYPERBOLA CENTERED AT ORIGIN

The general form of the equation of a hyperbola centered at the origin in the polar plane is given by the following formula:

$$r = ab/(a^2 \sin^2 \theta - b^2 \cos^2 \theta)^{1/2}$$

where a and b are real-number constants greater than 0.

Let D represent a rectangle whose center is at the origin, whose vertical edges are tangent to the hyperbola, and whose vertices (corners) lie on the *asymptotes* of the hyperbola (Fig. 5-6). Let a represent the distance from the origin to D as measured along the "horizontal" ray $\theta = 0$, and let b represent the distance from the origin to D as measured along the "vertical" ray $\theta = \pi/2$. The values a and b represent the lengths of the semi-axes of the hyperbola. The greater value is the length of the major semi-axis, and the lesser value is the length of the minor semi-axis.

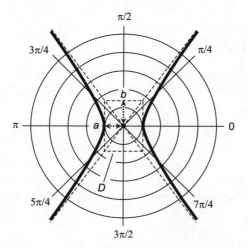

Fig. 5-6.   Polar graph of a hyperbola centered at the origin, with semi-axes a and b.

LEMNISCATE

The general form of the equation of a *lemniscate* centered at the origin in the polar plane is given by the following formula:

$$r = a(\cos 2\theta)^{1/2}$$

where a is a real-number constant greater than 0, representing the maximum radius. This is illustrated in Fig. 5-7.

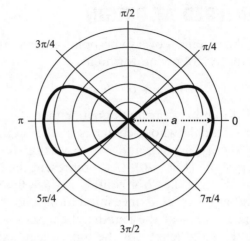

Fig. 5-7.   Polar graph of a lemniscate centered at the origin, with radius *a*.

THREE-LEAFED ROSE

The general form of the equation of a *three-leafed rose* centered at the origin in the polar plane is given by either of the following two formulas:

$$r = a \cos 3\theta$$
$$r = a \sin 3\theta$$

where *a* is a real-number constant greater than 0. The cosine curve is illustrated in Fig. 5-8A; the sine curve is illustrated in Fig. 5-8B.

FOUR-LEAFED ROSE

The general form of the equation of a *four-leafed rose* centered at the origin in the polar plane is given by either of the following two formulas:

$$r = a \cos 2\theta$$
$$r = a \sin 2\theta$$

where *a* is a real-number constant greater than 0. The cosine curve is illustrated in Fig. 5-9A; the sine curve is illustrated in Fig. 5-9B.

It is interesting, and a little bit mysterious, that the objects graphed in Figs. 5-8 and 5-9 are conventionally called "roses" and not "clovers."

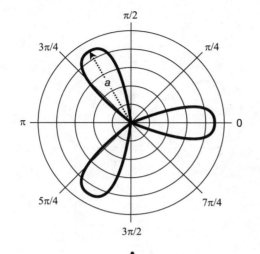

A

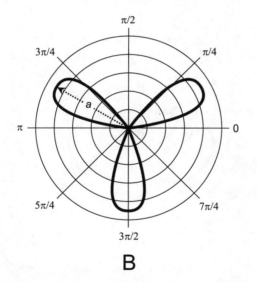

B

Fig. 5-8   (A) Polar graph of a three-leafed rose with equation $r = a \cos 3\theta$. (B) Polar graph of a three-leafed rose with equation $r = a \sin 3\theta$.

SPIRAL

The general form of the equation of a *spiral* centered at the origin in the polar plane is given by the following formula:

$$r = a\theta$$

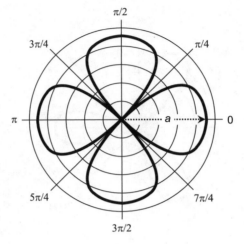

A

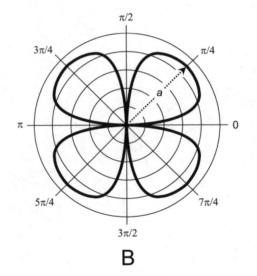

B

Fig. 5-9   (A) Polar graph of a four-leafed rose with equation $r = a \cos 2\theta$. (B) Polar graph of a four-leafed rose with equation $r = a \sin 2\theta$.

where a is a real-number constant greater than 0. An example of this type of spiral, called the *spiral of Archimedes* because of the uniform manner in which its radius increases as the angle increases, is illustrated in Fig. 5-10.

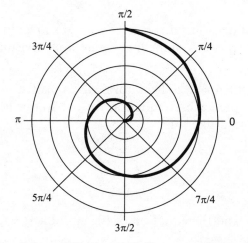

Fig. 5-10.   Polar graph of a spiral; illustration for Problem 5-1.

CARDIOID

The general form of the equation of a *cardioid* centered at the origin in the polar plane is given by the following formula:

$$r = 2a(1 + \cos\ \theta)$$

where *a* is a real-number constant greater than 0. An example of this type of curve is illustrated in Fig. 5-11.

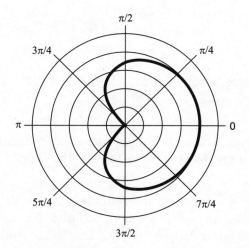

Fig. 5-11.   Polar graph of a cardioid; illustration for Problem 5-2.

PROBLEM 5-1
What is the value of the constant, a, in the spiral shown in Fig. 5-10? What is the equation of this spiral? Assume that each radial division represents 1 unit.

SOLUTION 5-1
Note that if $\theta = \pi$, then $r = 2$. Therefore, we can solve for a by substituting this number pair in the general equation for the spiral. Plug in the numbers $(\theta_0, r_0) = (\pi, 2)$. Proceed like this:

$$r_0 = a\theta_0$$
$$2 = a\pi$$
$$2/\pi = a$$

Therefore, $a = 2/\pi$, and the equation of the spiral is $r = (2/\pi)\theta$ or, in a form without parentheses, $r = 2\theta/\pi$.

PROBLEM 5-2
What is the value of the constant, a, in the cardioid shown in Fig. 5-11? What is the equation of this cardioid? Assume that each radial division represents 1 unit.

SOLUTION 5-2
Note that if $\theta = 0$, then $r = 4$. We can solve for a by substituting this number pair in the general equation for the cardioid. Plug in the numbers $(\theta_0, r_0) = (0, 4)$. Proceed like this:

$$r_0 = 2a(1 + \cos\ \theta_0)$$
$$4 = 2a(1 + \cos\ 0)$$
$$4 = 2a(1 + 1)$$
$$4 = 4a$$
$$a = 1$$

This means that the equation of the cardioid is $r = 2(1 + \cos\theta)$ or, in a form without parentheses, $r = 2 + 2\cos\theta$.

PROBLEM 5-3
What is the polar equation of a straight line running through the origin and ascending at a $45°$ angle as you move toward the right?

SOLUTION 5-3
The equation is $\theta = 45°$, or if we use radians, $\theta = \pi/4$. It is understood that the value of r can be any real number: positive, negative, or zero. If r is restricted to non-negative values, we get the closed-ended ray starting at the origin and pointing outward in the $45°$ ($\pi/4$ rad) direction. If r is restricted

to negative values, we get the open-ended ray starting at the origin and pointing outward in the 225° (5π/4 rad) direction. The union of these two rays forms the line running through the origin and ascending at a 45° angle as you move toward the right. In the rectangular xy-plane, this line is the graph of the equation $y = x$.

Compression and Conversion

Here are a couple of interesting things, one of which is presented as an exercise for the imagination, and the other of which has extensive applications in science and engineering.

GEOMETRIC POLAR PLANE

Figure 5-12 shows a variant of the polar coordinate plane on which the radial scale is graduated geometrically, rather than in linear fashion. The point corresponding to 1 on the r axis is halfway between the origin and the outer periphery, which is labeled ∞ (the "infinity" symbol). Succeeding integer points are placed halfway between previous integer points and the outer periphery. In this way, the entire polar coordinate plane is, in effect, portrayed inside an open circle having a finite radius.

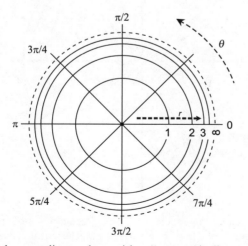

Fig. 5-12.   A polar coordinate plane with a "geometrically compressed" radial axis.

The radial scale of this coordinate system can be expanded or compressed by multiplying all the values on the r axis by a constant. This allows various relations and functions to be plotted, minimizing distortion in particular regions of interest. Distortion relative to the conventional polar coordinate plane is greatest near the periphery, and is least near the origin.

This "geometric axis compression" scheme can also be used with the axes of rectangular coordinates in two or three dimensions. It is not often seen in the literature, but it can be interesting because it provides a "view to infinity" that other coordinate systems do not.

MATHEMATICIAN'S POLAR VS CARTESIAN

Figure 5-13 shows a point $P = (x_0, y_0) = (\theta_0, r_0)$ graphed on superimposed Cartesian and polar coordinate systems. If we know the Cartesian coordinates, we can convert to polar coordinates using these formulas:

$$\theta_0 = \arctan (y_0/x_0) \text{ if } x_0 > 0$$
$$\theta_0 = 180° + \arctan (y_0/x_0) \text{ if } x_0 < 0 \text{ (for } \theta_0 \text{ in degrees)}$$

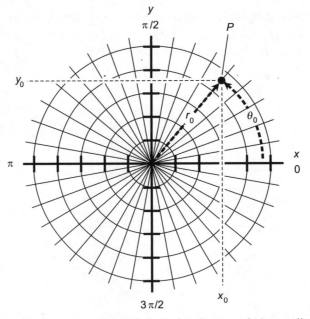

Fig. 5-13.   Conversion between polar and Cartesian (rectangular) coordinates. Each radial division represents one unit. Each division on the x and y axes also represents one unit.

$$\theta_0 = \pi + \arctan (y_0/x_0) \text{ if } x_0 < 0 \text{ (for } \theta_0 \text{ in radians)}$$
$$r_0 = (x_0^2 + y_0^2)^{1/2}$$

We can't have $x_0 = 0$ because that produces an undefined quotient in the conversion formula to θ_0. If a value of θ_0 thus determined happens to be negative, you can add 360° or 2π rad to get the "legitimate" value.

Polar coordinates are converted to Cartesian coordinates by the following formulas:

$$x_0 = r_0 \cos \theta_0$$
$$y_0 = r_0 \sin \theta_0$$

These same formulas can be used, by means of substitution, to convert Cartesian-coordinate relations to polar-coordinate relations, and vice versa. The general Cartesian-to-polar conversion formulas look like this:

$$\theta = \arctan (y/x) \text{ if } x > 0$$
$$\theta = 180° + \arctan (y/x) \text{ if } x < 0 \text{ (for } \theta \text{ in degrees)}$$
$$\theta = \pi + \arctan (y/x) \text{ if } x < 0 \text{ (for } \theta \text{ in radians)}$$
$$r = (x^2 + y^2)^{1/2}$$

The general polar-to-Cartesian conversion formulas are:

$$x = r \cos \theta$$
$$y = r \sin \theta$$

When making a conversion from polar to Cartesian coordinates or vice versa, a relation that is a function in one system is sometimes a function in the other system, but that is not always the case.

PROBLEM 5-4
Provide an example of a graphical object that can be represented as a function in polar coordinates, but not in Cartesian coordinates.

SOLUTION 5-4
In polar coordinates, let θ represent the independent variable, and let r represent the dependent variable. Then when we talk about a function f, we can say that $r = f(\theta)$. A simple function of θ in polar coordinates is a *constant function* such as this:

$$f(\theta) = 3$$

Because $f(\theta)$ is just another way of denoting r, the radius, this function tells us that $r = 3$. This is a circle with a radius of 3 units.

In Cartesian xy-coordinates, the equation of the circle with radius of 3 units is more complicated:

$$x^2 + y^2 = 9$$

(Note that $9 = 3^2$, the square of the radius.) If we let x be the independent variable and y be the dependent variable, we can rearrange the equation of the circle to get:

$$y = \pm(9 - x^2)^{1/2}$$

If we say that $y = g(x)$ where g is a function of x in this case, we are mistaken. There are values of x (the independent variable) that produce two values of y (the dependent variable). For example, when $x = 0$, $y = \pm 3$. If we want to say that g is a relation, that's fine, but we cannot call it a function.

PROBLEM 5-5
Consider the point $(\theta_0, r_0) = (135°, 2)$ in polar coordinates. What is the (x_0, y_0) representation of this point in Cartesian coordinates?

SOLUTION 5-5
Use the conversion formulas above:

$$x_0 = r_0 \cos \theta_0$$
$$y_0 = r_0 \sin \theta_0$$

Plugging in the numbers gives us these values, accurate to three decimal places:

$$x_0 = 2 \cos 135° = 2 \times (-0.707) = -1.414$$
$$y_0 = 2 \sin 135° = 2 \times 0.707 = 1.414$$

Thus, $(x_0, y_0) = (-1.414, 1.414)$.

The Navigator's Way

Navigators and military people use a coordinate plane similar to the one preferred by mathematicians. The radius is called the *range*, and real-world units are commonly specified, such as meters (m) or kilometers (km). The angle, or direction, is called the *azimuth*, *heading*, or *bearing*, and is measured in degrees clockwise from geographic north. The basic scheme is shown in Fig. 5-14. The azimuth is symbolized α (the lowercase Greek alpha), and the

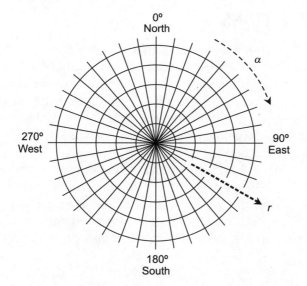

Fig. 5-14.   The navigator's polar coordinate plane. The azimuth, bearing, or heading α is in degrees measured clockwise from geographic north; the range r is in arbitrary units.

range is symbolized r. The position of a point is definable by an ordered pair (α, r).

WHAT IS NORTH?

There are two ways of defining "north," or $0°$. The more accurate, and thus the preferred and generally accepted, standard uses *geographic north*. This is the direction you should travel if you want to take the shortest possible route over the earth's surface to the north geographic pole. The less accurate standard uses *magnetic north*. This is the direction indicated by the needle in a magnetic compass.

For most locations on the earth's surface, there is a difference between geographic north and magnetic north. This difference, measured in degrees, is called *declination*. (This, by the way, is not the same thing as the declination used in celestial coordinates!) Navigators in olden times had to know the declination for their location when they couldn't use the stars to determine geographic north. Nowadays, there are electronic navigation systems such as the *Global Positioning System* (GPS) that are far more accurate than any magnetic compass, provided the equipment is in working order.

STRICT RESTRICTIONS

In navigator's polar coordinates, the range can never be negative. No navigator ever talks about traveling -20 km on a heading of $270°$, for example, when they really mean to say they are traveling 20 km on a heading of $90°$. When working out certain problems, it's possible that the result might contain a negative range. If this happens, the value of r should be multiplied by -1 and the value of α should be increased or decreased by $180°$ so the result is at least $0°$ but less than $360°$.

The azimuth, bearing, or heading must also conform to certain values. The smallest possible value of α is $0°$ (representing geographic north). As you turn clockwise as seen from above, the values of α increase through $90°$ (east), $180°$ (south), $270°$ (west), and ultimately approach, but never reach, $360°$ (north again). We therefore have these restrictions on the ordered pair (α,r):

$$0° \leq \alpha < 360°$$
$$r \geq 0$$

MATHEMATICIAN'S POLAR VS NAVIGATOR'S POLAR

Sometimes it is necessary to convert from mathematician's polar coordinates (let's call them MPC for short) to navigator's polar coordinates (NPC), or vice versa. When making the conversion, the radius of a particular point, r_0, is the same in both systems, so no change is necessary. But the angles differ.

If you know the direction angle θ_0 of a point in MPC and you want to find the equivalent azimuth α_0 in NPC, first be sure θ_0 is expressed in degrees, not radians. Then you can use either of the following conversion formulas, depending on the value of θ_0:

$$\alpha_0 = 90° - \theta_0 \text{ if } 0° \leq \theta_0 \leq 90°$$
$$\alpha_0 = 450° - \theta_0 \text{ if } 90° < \theta_0 < 360°$$

If you know the azimuth α_0 of a distant point in NPC and you want to find the equivalent direction angle θ_0 in MPC, then you can use either of the following conversion formulas, depending on the value of α_0:

$$\theta_0 = 90° - \alpha_0 \text{ if } 0° \leq \alpha_0 \leq 90°$$
$$\theta_0 = 450° - \alpha_0 \text{ if } 90° < \alpha_0 < 360°$$

NAVIGATOR'S POLAR VS CARTESIAN

Suppose you want to convert from NPC to Cartesian coordinates. Here are the conversion formulas for translating the coordinates for a point (α_0, r_0) in NPC to a point (x_0, y_0) in the Cartesian plane:

$$x_0 = r_0 \sin \alpha_0$$
$$y_0 = r_0 \cos \alpha_0$$

These are similar to the formulas used to convert MPC to Cartesian coordinates, except that the roles of the sine and cosine function are reversed.

In order to convert the coordinates of a point (x_0, y_0) in Cartesian coordinates to a point (α_0, r_0) in NPC, use these formulas:

$$\alpha_0 = \arctan (x_0/y_0) \text{ if } y_0 > 0$$
$$\alpha_0 = 180° + \arctan (x_0/y_0) \text{ if } y_0 < 0$$
$$r_0 = (x_0^2 + y_0^2)^{1/2}$$

We can't have $y_0 = 0$, because that produces an undefined quotient. If a value of α_0 thus determined happens to be negative, add 360° to get the "legitimate" value. These are similar to the formulas used to convert Cartesian coordinates to MPC.

PROBLEM 5-6
Suppose a radar set with an NPC display indicates the presence of a hovering object at a bearing of 300° and a range of 40 km. If we say that a kilometer is the same as a "unit," what are the coordinates (θ_0, r_0) of this object in MPC? Express θ_0 in both degrees and radians.

SOLUTION 5-6
We are given coordinates $(\alpha_0, r_0) = (300°, 40)$. The value of r_0, the radius, is the same as the range, in this case 40 units. As for the angle θ_0, remember the conversion formulas given above. In this case, α_0 is greater than 90° and less than 360°. Therefore:

$$\theta_0 = 450° - \alpha_0$$
$$\theta_0 = 450° - 300° = 150°$$

Therefore, $(\theta_0, r_0) = (150°, 40)$. To express θ_0 in radians, recall that there are 2π radians in a full 360° circle, or π radians in a 180° angle. Note that 150° is exactly 5/6 of 180°. Therefore, $\theta_0 = 5\pi/6$ rad, and we can say that $(\theta_0, r_0) = (150°, 40) = (5\pi/6, 40)$. We can leave the "rad" off the angle designator here, because when units are not specified for an angle, radians are assumed.

PROBLEM 5-7

Suppose you are on an archeological expedition, and you unearth a stone on which appears a treasure map. The map says "You are here" next to an X, and then says, "Go north 40 paces and then west 30 paces." Suppose that you let west represent the negative x axis of a Cartesian coordinate system, east represent the positive x axis, south represent the negative y axis, and north represent the positive y axis. Also suppose that you let one "pace" represent one "unit" of radius, and also one "unit" in the Cartesian system. If you are naïve enough to look for the treasure and lazy enough so you insist on walking in a straight line to reach it, how many paces should you travel, and in what direction, in NPC? Determine your answer to the nearest degree, and to the nearest pace.

SOLUTION 5-7

First, determine the ordered pair in Cartesian coordinates that corresponds to the imagined treasure site. Consider the origin to be the spot where the map was unearthed. If we let (x_0, y_0) be the point where the treasure should be, then 40 paces north means $y_0 = 40$, and 30 paces west means $x_0 = -30$:

$$(x_0, y_0) = (-30, 40)$$

Because y_0 is positive, we use this formula to determine the bearing or heading α_0:

$$\alpha_0 = \arctan (x_0/y_0)$$
$$= \arctan (-30/40)$$
$$= \arctan -0.75$$
$$= -37°$$

This is a negative angle, so to get it into the standard form, we must add $360°$:

$$\alpha_0 = -37° + 360° = 360° - 37°$$
$$= 323°$$

To find the value of the range, r_0, use this formula:

$$r_0 = (x_0^2 + y_0^2)^{1/2}$$
$$= (30^2 + 40^2)^{1/2}$$
$$= (900 + 1600)^{1/2}$$
$$= 2500^{1/2}$$
$$= 50$$

This means $(\alpha_0, r_0) = (323°, 50)$. Proceed 50 paces at a heading of $323°$ (approximately north by northwest). Then, if you wish, go ahead and dig!

Quiz

Refer to the text in this chapter if necessary. A good score is eight correct. Answers are in the back of the book.

1. The equal-radius axes in the mathematician's polar coordinate system are
   (a) rays
   (b) lines
   (c) circles
   (d) spirals

2. Suppose a point has the coordinates $(\theta, r) = (\pi, 3)$ in the mathematician's polar scheme. It is implied from this that the angle is
   (a) negative
   (b) expressed in radians
   (c) greater than $360°$
   (d) ambiguous

3. Suppose a point has the coordinates $(\theta, r) = (\pi/4, 6)$ in the mathematician's polar scheme. What are the coordinates (α, r) of the point in the navigator's polar scheme?
   (a) They cannot be determined without more information
   (b) $(-45°, 6)$
   (c) $(45°, 6)$
   (d) $(135°, 6)$

4. Suppose we are given the simple relation $g(x) = x$. In Cartesian coordinates, this has the graph $y = x$. What is the equation that represents the graph of this relation in the mathematician's polar coordinate system?
   (a) $r = \theta$
   (b) $r = 1/\theta$, where $\theta \neq 0°$
   (c) $\theta = 45°$, where r can range over the entire set of real numbers
   (d) $\theta = 45°$, where r can range over the set of non-negative real numbers

5. Suppose we set off on a bearing of 135° in the navigator's polar coordinate system. We stay on a straight course. If the starting point is considered the origin, what is the graph of our path in Cartesian coordinates?
   (a) $y = x$, where $x \geq 0$
   (b) $y = 0$, where $x \geq 0$
   (c) $x = 0$, where $y \geq 0$
   (d) $y = -x$, where $x \geq 0$

6. The direction angle in the navigator's polar coordinate system is measured
   (a) in a clockwise sense
   (b) in a counterclockwise sense
   (c) in either sense
   (d) only in radians

7. The graph of $r = -3\theta$ in the mathematician's polar coordinate system looks like
   (a) a circle
   (b) a cardioid
   (c) a spiral
   (d) nothing; it is undefined

8. A function in polar coordinates
   (a) is always a function in rectangular coordinates
   (b) is sometimes a function in rectangular coordinates
   (c) is never a function in rectangular coordinates
   (d) cannot have a graph that is a straight line

9. Suppose we are given a point and told that its Cartesian coordinate is $(x,y) = (0,-5)$. In the mathematician's polar scheme, the coordinates of this point are
   (a) $(\theta,r) = (3\pi/2,5)$
   (b) $(\theta,r) = (3\pi/2,-5)$
   (c) $(\theta,r) = (-5,3\pi/2)$
   (d) ambiguous; we need more information to specify them

10. Suppose a radar unit shows a target that is 10 kilometers away in a southwesterly direction. It is moving directly away from us. When its distance has doubled to 20 kilometers, what has happened to the x and y coordinates of the target in Cartesian coordinates? Assume we are located at the origin.

(a) They have both doubled
(b) They have both increased by a factor equal to the square root of 2
(c) They have both quadrupled
(d) We need to specify the size of each unit in the Cartesian coordinate system in order to answer this question

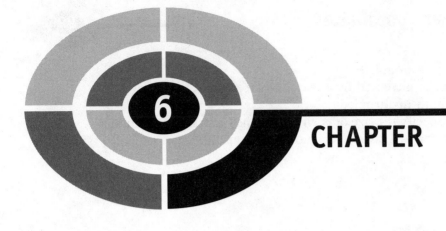

CHAPTER

Three-Space and Vectors

It's time for us to get into a three-dimensional (3D) frame of mind. In this chapter, we leave the flatness and simplicity of the two-dimensional (2D) plane, and venture into space where things can go north, south, east, west, up, or down.

Spatial Coordinates

Here are some coordinate systems that are used in mathematics and science when working in 3D space.

LATITUDE AND LONGITUDE

Latitude and *longitude* are directional angles that uniquely define the positions of points on the surface of a sphere or in the sky. The scheme for geographic locations on the earth is illustrated in Fig. 6-1A. The *polar axis* connects two specified points at *antipodes*, or points directly opposite each other, on the sphere. These points are assigned latitude $\theta = 90°$ (north pole) and $\theta = -90°$ (south pole). The *equatorial axis* runs outward from the center of the sphere at a right angle to the polar axis. It is assigned longitude $\phi = 0°$.

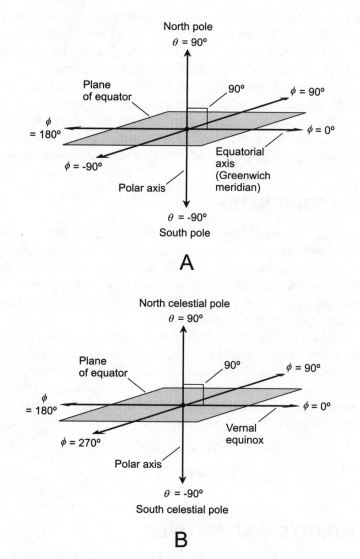

Fig. 6-1. (A) Latitude and longitude angles for locating points on the earth's surface. (B) Declination and right ascension angles for locating points in the sky.

Latitude θ is measured positively (north) and negatively (south) relative to the plane of the equator. Longitude ϕ is measured counterclockwise (positively) and clockwise (negatively) relative to the equatorial axis. The angles are restricted as follows:

$$-90° \leq \theta \leq 90°$$
$$-180° < \phi \leq 180°$$

On the earth's surface, the half-circle connecting the 0° longitude line with the poles passes through Greenwich, England (not Greenwich Village in New York City!) and is known as the *Greenwich meridian* or the *prime meridian*. Longitude angles are defined with respect to this meridian.

Latitude and longitude angles translate into points on the surface of a sphere (such as the earth's surface) centered at the point where the polar axis intersects the equatorial plane. But latitude and longitude angles can also translate into positions in the sky. These positions are not really points, but are rays pointing out from the observer's eyes indefinitely into space.

CELESTIAL COORDINATES

Celestial latitude and *celestial longitude* are extensions of the earth's latitude and longitude angles into the heavens. The same set of coordinates used for geographic latitude and longitude applies to this system. An object whose celestial latitude and longitude coordinates are (θ,ϕ) appears at the *zenith* in the sky, that is, directly overhead, from the point on the earth's surface whose latitude and longitude coordinates are (θ,ϕ).

Declination and *right ascension* define the positions of objects in the sky relative to the stars, rather than the earth. Figure 6-1B applies to this system. Declination (θ) is identical to celestial latitude. Right ascension (ϕ) is measured eastward from the *vernal equinox*, which is the position of the sun in the heavens at the moment spring begins in the northern hemisphere. The angles are restricted as follows:

$$-90° \leq \theta \leq 90°$$
$$0° \leq \phi < 360°$$

HOURS, MINUTES, AND SECONDS

Astronomers use a peculiar scheme for right ascension. Instead of expressing the angles of right ascension in degrees or radians, they use *hours*, *minutes*, and *seconds* based on 24 hours in a complete circle, corresponding to the 24 hours in a day. That means each hour of right ascension is equivalent to 15°.

If that isn't confusing enough for you, the minutes and seconds of right ascension are not the same as the fractional degree units by the same names that are used by mathematicians and engineers. One minute of right ascension is 1/60 of an hour or $\frac{1}{4}$ of an angular degree, and one second of right ascension is 1/60 of a minute or 1/240 of an angular degree.

CARTESIAN THREE-SPACE

An extension of rectangular coordinates into three dimensions is *Cartesian three-space* (Fig. 6-2), also called *xyz-space*. The independent variables are usually plotted along the x and y axes; the dependent variable is plotted along the z axis. Each axis is perpendicular to the other two. They all intersect at the *origin*, which is usually the point where $x = 0$, $y = 0$, and $z = 0$.

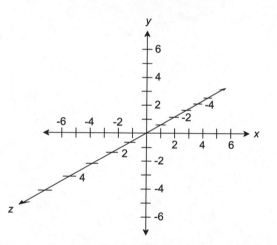

Fig. 6-2.   Cartesian three-space, also called *xyz*-space.

The scales in Cartesian three-space are all linear. This means that, along any given individual axis, equal distances represent equal changes in value. But the divisions (that is, the spaces between hash marks) on different axes do not necessarily have to represent the same increments. For example, the x axis might be designated as having 1 unit per division, the y axis 10 units per division, and the z axis 5 units per division.

Points in Cartesian three-space are represented by ordered triples (x,y,z). As with ordered pairs, there are no spaces between the variables and the commas when denoting an ordered triple; they're all scrunched up together.

CYLINDRICAL COORDINATES

Figure 6-3 shows two systems of *cylindrical coordinates* for specifying the positions of points in three-space.

In the system shown in Fig. 6-3A, we start with Cartesian *xyz*-space. Then an angle θ is defined in the *xy*-plane, measured in degrees or radians (usually radians) counterclockwise from the positive *x* axis or *reference axis*. Given a point *P* in space, consider its projection *P'* onto the *xy*-plane, such that line

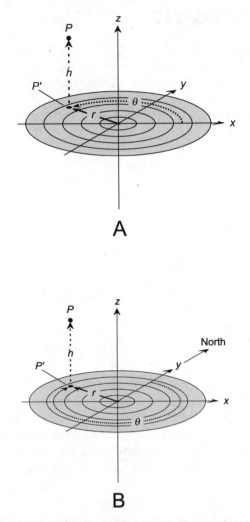

A

B

Fig. 6-3. (A) Mathematician's form of cylindrical coordinates for defining points in three-space. (B) Astronomer's and navigator's form of cylindrical coordinates for defining points in three-space.

segment PP' is parallel to the z axis. The position of P is defined by the ordered triple (θ, r, h). In this ordered triple, θ represents the angle measured counterclockwise between P' and the positive x axis in the xy-plane, r represents the distance or radius from P' to the origin, and h represents the distance (altitude or height) of P above the xy-plane. This scheme for cylindrical coordinates is preferred by mathematicians, and also by some engineers and scientists.

In the system shown in Fig. 6-3B, we again start with Cartesian xyz-space. The xy-plane corresponds to the surface of the earth in the vicinity of the origin, and the z axis runs straight up (positive z values) and down (negative z values). The angle θ is defined in the xy-plane in degrees (but never radians) *clockwise* from the positive y axis, which corresponds to geographic north. Given a point P in space, consider its projection P' onto the xy-plane, such that line segment PP' is parallel to the z axis. The position of P is defined by the ordered triple (θ, r, h), where θ represents the angle measured clockwise between P' and geographic north, r represents the distance or radius from P' to the origin, and h represents the distance (altitude or height) of P above the xy-plane. This scheme is preferred by navigators and astronomers.

SPHERICAL COORDINATES

Figure 6-4 shows three systems of *spherical coordinates* for defining points in space. The first two are used by astronomers and aerospace scientists, while the third one is preferred by navigators and surveyors.

In the scheme shown in Fig. 6-4A, the location of a point P is defined by the ordered triple (θ, ϕ, r) such that θ represents the declination of P, ϕ represents the right ascension of P, and r represents the radius from P to the origin, also called the *range*. In this example, angles are specified in degrees (except in the case of the astronomer's version of right ascension, which is expressed in hours, minutes, and seconds as defined earlier in this chapter). Alternatively, the angles can be expressed in radians. This system is fixed relative to the stars.

Instead of declination and right ascension, the variables θ and ϕ can represent celestial latitude and celestial longitude respectively, as shown in Fig. 6-4B. This system is fixed relative to the earth, rather than relative to the stars.

There's yet another alternative: θ can represent elevation (the angle above the horizon) and ϕ can represent the azimuth (bearing or heading), measured clockwise from geographic north. In this case, the reference plane corresponds to the horizon, not the equator, and the elevation can cover the

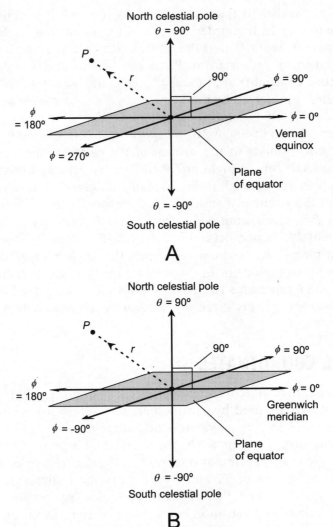

Fig. 6-4. (A) Spherical coordinates for defining points in three-space, where the angles θ and ϕ represent declination and right ascension, and r represents radius or range. (B) Spherical coordinates for defining points in three-space, where the angles θ and ϕ represent latitude and longitude, and r represents radius or range.

span of values between, and including, $-90°$ (the *nadir*, or the point directly underfoot) and $+90°$ (the zenith). This is shown in Fig. 6-4C. In a variant of this system used by mathematicians, the angle θ is measured with respect to the zenith (or the positive z axis), rather than the plane of the horizon. Then the range for this angle is $0° \leq \theta \leq 180°$.

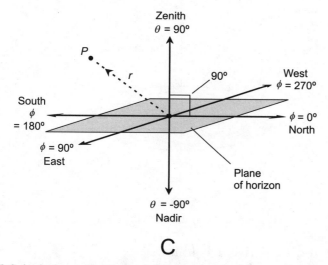

Fig. 6-4.   (C) Spherical coordinates for defining points in three-space, where the angles θ and ϕ represent elevation (angle above the horizon) and azimuth (also called bearing or heading), and r represents radius or range.

PROBLEM 6-1

Suppose you fly a kite above a perfectly flat, level field. The wind is out of the east–southeast, or azimuth 120°. Thus, the kite flies in a west–northwesterly direction, at azimuth 300°. Suppose the kite flies at an elevation angle of 50° above the horizon, and the kite line is 100 meters long. Imagine that it is a sunny day, and the sun is exactly overhead, so the kite's shadow falls directly underneath it. How far from you is the shadow of the kite? How high is the kite above the ground? Express your answers to the nearest meter.

SOLUTION 6-1

Let's work in navigator's cylindrical coordinates. The important factors are the length of the kite line (100 meters) and the angle at which the kite flies (50°). Figure 6-5 shows the scenario. Let r be the distance of the shadow from you, as expressed in meters. Let h be the height of the kite above the ground, also in meters.

First, let's find the ratio of h to the length of the kite line, that is, $h/100$. The line segment whose length is h, the line segment whose length is r, and the kite line form a right triangle with the hypotenuse corresponding to the kite line. From basic circular trigonometry, we can surmise the following:

$$\sin 50° = h/100$$

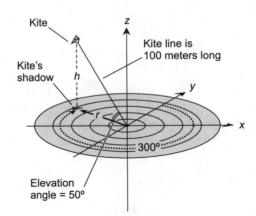

Fig. 6-5.   Illustration for Problem 6-1.

Using a calculator, we derive h as follows:

$$\sin\ 50° = 0.766 = h/100$$
$$76.6 = h$$
$$h = 77 \text{ meters (rounded off to nearest meter)}$$

We also know, from basic circular trigonometry, this:

$$\cos\ 50° = r/100$$

Using a calculator, we derive r as follows:

$$\cos\ 50° = 0.643 = r/100$$
$$64.3 = r$$
$$r = 64 \text{ meters (rounded off to nearest meter)}$$

In this situation, the wind direction is irrelevant. But if the sun were not directly overhead, the wind direction would make a difference. It would also make the problem a lot more complicated. If you like difficult problems, try this one again, but imagine that the sun is shining from the southern sky (azimuth 180°) and is at an angle of 35° above the horizon.

Vectors in the Cartesian Plane

A *vector* is a mathematical expression for a quantity with two independent properties: *magnitude* and *direction*. Vectors are used to represent physical variables such as displacement, velocity, and acceleration, when such variables have both magnitude and direction.

Conventionally, vectors are denoted by boldface letters of the alphabet. In the xy-plane, vectors **a** and **b** can be illustrated as rays from the origin (0,0) to points (x_a, y_a) and (x_b, y_b) as shown in Fig. 6-6.

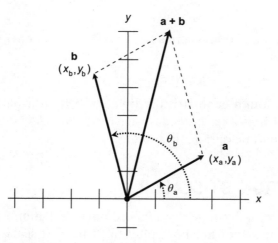

Fig. 6-6.   Two vectors in the Cartesian plane. They are added using the "parallelogram method."

MAGNITUDE

The magnitude, or length, of a vector **a**, written $|\mathbf{a}|$ or a, can be found in the Cartesian plane by using a distance formula resembling the Pythagorean theorem:

$$|\mathbf{a}| = (x_a^2 + y_a^2)^{1/2}$$

DIRECTION

The direction of **a**, written dir **a**, is the angle θ_a that **a** subtends counterclockwise from the positive x axis. This angle is equal to the arctangent of the ratio of y_a to x_a:

$$\text{dir } \mathbf{a} = \theta_a = \arctan (y_a/x_a)$$

By convention, the following restrictions hold:

$$0 \leq \theta_a < 360 \text{ for } \theta_a \text{ in degrees}$$
$$0 \leq \theta_a < 2\pi \text{ for } \theta_a \text{ in radians}$$

SUM

The sum of vectors **a** and **b**, where $\mathbf{a} = (x_a, y_a)$ and $\mathbf{b} = (x_b, y_b)$, is given by the following formula:

$$\mathbf{a} + \mathbf{b} = [(x_a + x_b), (y_a + y_b)]$$

This sum can be found geometrically by constructing a parallelogram with the vectors **a** and **b** as adjacent sides; the vector **a** + **b** is determined by the diagonal of this parallelogram (Fig. 6-6).

MULTIPLICATION BY SCALAR

To multiply a vector by a *scalar* (an ordinary real number), the *x* and *y* components of the vector are both multiplied by that scalar. Multiplication by a scalar is commutative. This means that it doesn't matter whether the scalar comes before or after the vector in the product. If we have a vector $\mathbf{a} = (x_a, y_a)$ and a scalar k, then

$$k\mathbf{a} = \mathbf{a}k = (kx_a, ky_a)$$

DOT PRODUCT

Let $\mathbf{a} = (x_a, y_a)$ and $\mathbf{b} = (x_b, y_b)$. The *dot product*, also known as the *scalar product* and written **a** · **b**, of vectors **a** and **b** is a real number (that is, a scalar) given by the formula:

$$\mathbf{a} \cdot \mathbf{b} = x_a x_b + y_a y_b$$

PROBLEM 6-2
What is the sum of $\mathbf{a} = (3, -5)$ and $\mathbf{b} = (2, 6)$?

SOLUTION 6-2
Add the *x* and *y* components together independently:

$$\mathbf{a} + \mathbf{b} = [(3 + 2), (-5 + 6)]$$
$$= (5, 1)$$

PROBLEM 6-3
What is the dot product of $\mathbf{a} = (3,-5)$ and $\mathbf{b} = (2,6)$?

SOLUTION 6-3
Use the formula given above for the dot product:

$$\mathbf{a} \cdot \mathbf{b} = (3 \times 2) + (-5 \times 6)$$
$$= 6 + (-30)$$
$$= -24$$

PROBLEM 6-4
What happens if the order of the dot product is reversed? Does the value change?

SOLUTION 6-4
No. The dot product of two vectors does not depend on the order in which the vectors are "dot-multiplied." This can be proven in the general case using the formula above. Let $\mathbf{a} = (x_a, y_a)$ and $\mathbf{b} = (x_b, y_b)$. First consider the dot product of $\mathbf{a}$ and $\mathbf{b}$ (pronounced "$\mathbf{a}$ dot $\mathbf{b}$"):

$$\mathbf{a} \cdot \mathbf{b} = x_a x_b + y_a y_b$$

Now consider the dot product $\mathbf{b} \cdot \mathbf{a}$:

$$\mathbf{b} \cdot \mathbf{a} = x_b x_a + y_b y_a$$

Because multiplication is commutative for all real numbers, the above formula is equivalent to:

$$\mathbf{b} \cdot \mathbf{a} = x_a x_b + y_a y_b$$

But $x_a x_b + y_a y_b$ is the expansion of $\mathbf{a} \cdot \mathbf{b}$. Therefore, for any two vectors $\mathbf{a}$ and $\mathbf{b}$, it is always true that $\mathbf{a} \cdot \mathbf{b} = \mathbf{b} \cdot \mathbf{a}$.

Vectors in the Polar Plane

In the mathematician's polar coordinate plane, vectors $\mathbf{a}$ and $\mathbf{b}$ can be denoted as rays from the origin $(0,0)$ to points (θ_a, r_a) and (θ_b, r_b) as shown in Fig. 6-7.

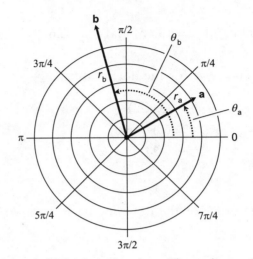

Fig. 6-7. Two vectors **a** and **b** in the polar plane. The angles are θ_a and θ_b. All angles are expressed in radians. The radii are r_a and r_b.

MAGNITUDE AND DIRECTION

The magnitude and direction of vector **a** in the polar coordinate plane are defined directly:

$$|\mathbf{a}| = r_a$$
$$\text{dir } \mathbf{a} = \theta_a$$

By convention, the following restrictions hold:

$$0° \leq \theta_a < 360° \text{ for } \theta_a \text{ in degrees}$$
$$0 \leq \theta_a < 2\pi \text{ for } \theta_a \text{ in radians}$$
$$r_a \geq 0$$

SUM

The sum of two vectors **a** and **b** in polar coordinates is best found by converting them into their equivalents in rectangular (xy-plane) coordinates, adding the vectors according to the formula for the xy-plane, and then converting the resultant back to polar coordinates. To convert vector **a** from polar to rectangular coordinates, these formulas apply:

$$x_a = r_a \cos \theta_a$$
$$y_a = r_a \sin \theta_a$$

To convert vector **a** from rectangular coordinates to polar coordinates, these formulas apply:

$$\theta_a = \arctan (y_a/x_a) \text{ if } x_a > 0$$
$$\theta_a = 180° + \arctan (y_a/x_a) \text{ if } x_a < 0 \text{ (for } \theta_a \text{ in degrees)}$$
$$\theta_a = \pi + \arctan (y_a/x_a) \text{ if } x_a < 0 \text{ (for } \theta_a \text{ in radians)}$$
$$r_a = (x_a^2 + y_a^2)^{1/2}$$

MULTIPLICATION BY SCALAR

In two-dimensional polar coordinates, let vector **a** be defined by the coordinates (θ,r) as shown in Fig. 6-8. Suppose **a** is multiplied by a positive real scalar k. Then the following equation holds:

$$k\mathbf{a} = (\theta,kr)$$

If **a** is multiplied by a negative real scalar $-k$, then:

$$-k\mathbf{a} = [(\theta + 180°), kr]$$

for θ in degrees. For θ in radians, the formula is:

$$-k\mathbf{a} = [(\theta + \pi), kr]$$

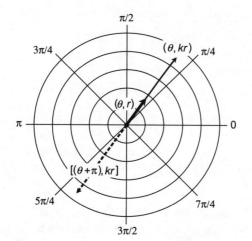

Fig. 6-8. Multiplication of a polar-plane vector **a** by a positive real scalar k, and by a negative real scalar $-k$. All angles are expressed in radians.

The addition of $180°$ (π rad) to θ reverses the direction of **a**. The same effect can be produced by subtracting $180°$ (π rad) from θ.

DOT PRODUCT

Let r_a be the radius of vector **a**, and r_b be the radius of vector **b** in the polar plane. Then the dot product of **a** and **b** is given by:

$$\mathbf{a} \cdot \mathbf{b} = |a||b| \cos (\theta_b - \theta_a)$$
$$= r_a r_b \cos (\theta_b - \theta_a)$$

PROBLEM 6-5

Consider the vector $\mathbf{a}_c = (x_a, y_a) = (3,4)$ in Cartesian coordinates. What is the equivalent vector $\mathbf{a}_p = (\theta_a, r_a)$ in mathematician's polar coordinates? Express values to the nearest hundredth of a linear unit, and to the nearest degree.

SOLUTION 6-5

Use the conversion formulas above. First find the direction angle θ_a. Because $x_a > 0$, we use this formula:

$$\theta_a = \arctan (y_a/x_a)$$
$$= \arctan (4/3)$$
$$= \arctan 1.333$$
$$= 53°$$

Solving for r_a, we proceed as follows:

$$r_a = (x_a^2 + y_a^2)^{1/2}$$
$$= (3^2 + 4^2)^{1/2}$$
$$= (9 + 16)^{1/2}$$
$$= 25^{1/2}$$
$$= 5.00$$

Therefore, $\mathbf{a}_p = (\theta_a, r_a) = (53°, 5.00)$.

PROBLEM 6-6

Consider the vector $\mathbf{b}_p = (\theta_b, r_b) = (200°, 4.55)$ in mathematician's polar coordinates. Convert this to an equivalent vector $\mathbf{b}_c = (x_b, y_b)$ in Cartesian coordinates. Express your answer to the nearest tenth of a unit for both coordinates x_b and y_b.

SOLUTION 6-6

Use the conversion formulas above. First, solve for x_b:

$$x_b = r_b \ \cos \ \theta_b$$
$$= 4.55 \ \cos \ 200°$$
$$= 4.55 \times (-0.9397)$$
$$= -4.3$$

Next, solve for y_b:

$$y_b = r_b \ \sin \ \theta_b$$
$$= 4.55 \ \sin \ 200°$$
$$= 4.55 \times (-0.3420)$$
$$= -1.6$$

Therefore, $\mathbf{b}_c = (x_b, y_b) = (-4.3, -1.6)$.

Vectors in 3D

In rectangular xyz-space, vectors $\mathbf{a}$ and $\mathbf{b}$ can be denoted as rays from the origin $(0,0,0)$ to points (x_a, y_a, z_a) and (x_b, y_b, z_b) as shown in Fig. 6-9.

MAGNITUDE

The magnitude of $\mathbf{a}$, written $|\mathbf{a}|$, can be found by a three-dimensional extension of the Pythagorean theorem for right triangles. The formula looks like this:

$$|\mathbf{a}| = (x_a^2 + y_a^2 + z_a^2)^{1/2}$$

DIRECTION

The direction of $\mathbf{a}$ is denoted by measuring the angles θ_x, θ_y, and θ_z that the vector $\mathbf{a}$ subtends relative to the positive x, y, and z axes respectively (Fig. 6-10). These angles, expressed in radians as an ordered triple $(\theta_x, \theta_y, \theta_z)$, are

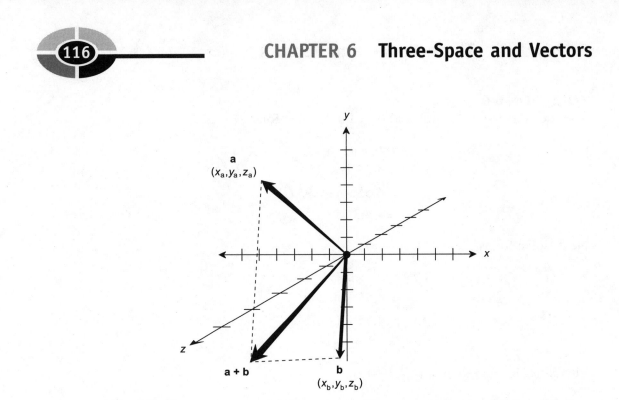

Fig. 6-9. Two vectors **a** and **b** in *xyz*-space. They are added using the "parallelogram method." This is a perspective drawing, so the parallelogram appears distorted.

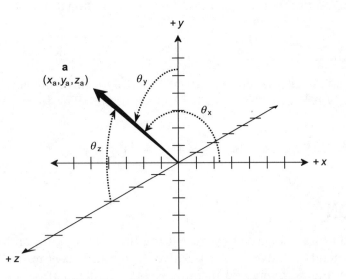

Fig. 6-10. Direction angles of a vector in *xyz*-space.

the *direction angles* of **a**. Sometimes the cosines of these angles are specified. These are the *direction cosines* of **a**:

$$\text{dir } \mathbf{a} = (\alpha, \beta, \gamma)$$
$$\alpha = \cos \theta_x$$
$$\beta = \cos \theta_y$$
$$\gamma = \cos \theta_z$$

SUM

The sum of vectors **a** and **b** is:

$$\mathbf{a} + \mathbf{b} = [(x_a + x_b), (y_a + y_b), (z_a + z_b)]$$

This sum can, as in the two-dimensional case, be found geometrically by constructing a parallelogram with **a** and **b** as adjacent sides. The sum **a** + **b** is determined by the diagonal of the parallelogram, as shown in Fig. 6-9.

MULTIPLICATION BY SCALAR

In three-dimensional Cartesian coordinates, let vector **a** be defined by the coordinates (x_a, y_a, z_a). Suppose **a** is multiplied by some positive real scalar k. Then the following equation holds:

$$k\mathbf{a} = k(x_a, y_a, z_a) = (kx_a, ky_a, kz_a)$$

If **a** is multiplied by a negative real scalar $-k$, then:

$$-k\mathbf{a} = -k(x_a, y_a, z_a) = (-kx_a, -ky_a, -kz_a)$$

Suppose the direction angles of **a** are represented by $(\theta_x, \theta_y, \theta_z)$. The direction angles of $k\mathbf{a}$ are also $(\theta_x, \theta_y, \theta_z)$. The direction angles of $-k\mathbf{a}$ are all increased by 180° (π rad), so they are represented by $[(\theta_x + \pi), (\theta_y + \pi), (\theta_z + \pi)]$. The same effect can be accomplished by subtracting 180° (π rad) from each of these direction angles.

DOT PRODUCT

The *dot product*, also known as the *scalar product* and written **a** · **b**, of vectors **a** and **b** in Cartesian *xyz*-space is a real number given by the formula:

$$\mathbf{a} \cdot \mathbf{b} = x_a x_b + y_a y_b + z_a z_b$$

where $\mathbf{a} = (x_a, y_a, z_a)$ and $\mathbf{b} = (x_b, y_b, z_b)$.

The dot product $\mathbf{a} \cdot \mathbf{b}$ can also be found from the magnitudes $|\mathbf{a}|$ and $|\mathbf{b}|$, and the angle θ between vectors $\mathbf{a}$ and $\mathbf{b}$ as measured counterclockwise in the plane containing them both:

$$\mathbf{a} \cdot \mathbf{b} = |\mathbf{a}||\mathbf{b}| \cos \theta$$

CROSS PRODUCT

The *cross product*, also known as the *vector product* and written $\mathbf{a} \times \mathbf{b}$, of vectors $\mathbf{a}$ and $\mathbf{b}$ is a vector perpendicular to the plane containing $\mathbf{a}$ and $\mathbf{b}$. Let θ be the angle between vectors $\mathbf{a}$ and $\mathbf{b}$ expressed counterclockwise (as viewed from above, or the direction of the positive z axis) in the plane containing them both (Fig. 6-11). The magnitude of $\mathbf{a} \times \mathbf{b}$ is given by the formula:

$$|\mathbf{a} \times \mathbf{b}| = |\mathbf{a}||\mathbf{b}| \sin \theta$$

In the example shown, $\mathbf{a} \times \mathbf{b}$ points upward at a right angle to the plane containing both vectors $\mathbf{a}$ and $\mathbf{b}$. If $0° < \theta < 180°$ ($0 < \theta < \pi$), you can use the *right-hand rule* to ascertain the direction of $\mathbf{a} \times \mathbf{b}$. Curl your fingers in the sense in which θ, the angle between $\mathbf{a}$ and $\mathbf{b}$, is defined. Extend your thumb. Then $\mathbf{a} \times \mathbf{b}$ points in the direction of your thumb.

When $180° < \theta < 360°$ ($\pi < \theta < 2\pi$), the cross-product vector reverses direction compared with the situation when $0° < \theta < 180°$ ($0 < \theta < \pi$). This is demonstrated by the fact that, in the above formula, $\sin \theta$ is positive when $0° < \theta < 180°$ ($0 < \theta < \pi$), but negative when $180° < \theta < 360°$ ($\pi < \theta < 2\pi$). When $180° < \theta < 360°$ ($\pi < \theta < 2\pi$), the right-hand rule doesn't work.

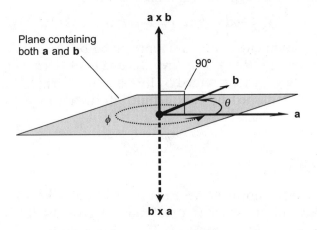

Fig. 6-11.   The vector $\mathbf{b} \times \mathbf{a}$ has the same magnitude as vector $\mathbf{a} \times \mathbf{b}$, but points in the opposite direction. Both vectors $\mathbf{b} \times \mathbf{a}$ and $\mathbf{a} \times \mathbf{b}$ are perpendicular to the plane defined by $\mathbf{a}$ and $\mathbf{b}$.

Instead, you must use your left hand, and curl your fingers into almost a complete circle! An example is the cross product $\mathbf{b} \times \mathbf{a}$ in Fig. 6-11. The angle ϕ, expressed counterclockwise between these vectors (as viewed from above), is more than 180°.

For any two vectors $\mathbf{a}$ and $\mathbf{b}$, the vector $\mathbf{b} \times \mathbf{a}$ is a "mirror image" of $\mathbf{a} \times \mathbf{b}$, where the "mirror" is the plane containing both vectors. One way to imagine the "mirror image" is to consider that $\mathbf{b} \times \mathbf{a}$ has the same magnitude as $\mathbf{a} \times \mathbf{b}$, but points in exactly the opposite direction. Putting it another way, the direction of $\mathbf{b} \times \mathbf{a}$ is the same as the direction of $\mathbf{a} \times \mathbf{b}$, but the magnitudes of the two vectors are additive inverses (negatives of each other). The cross product operation is not commutative, but the following relationship holds:

$$\mathbf{a} \times \mathbf{b} = -(\mathbf{b} \times \mathbf{a})$$

A POINT OF CONFUSION

Are you confused here about the concept of vector magnitude, and the fact that absolute-value symbols (the two vertical lines) are used to denote magnitude? The absolute value of a number is always positive, but with vectors, negative magnitudes sometimes appear in the equations.

Whenever we see a vector whose magnitude is negative, it is the equivalent of a positive vector pointing in the opposite direction. For example, if a force of -20 newtons is exerted upward, it is the equivalent of a force of 20 newtons exerted downward. When a vector with negative magnitude occurs in the final answer to a problem, you can reverse the direction of the vector, and assign to it a positive magnitude that is equal to the absolute value of the negative magnitude.

PROBLEM 6-7
What is the magnitude of the vector denoted by $\mathbf{a} = (x_a, y_a, z_a) = (1,2,3)$? Consider the values 1, 2, and 3 to be exact; express the answer to four decimal places.

SOLUTION 6-7
Use the distance formula for a vector in Cartesian xyz-space:

$$\begin{aligned}
|\mathbf{a}| &= (x_a^2 + y_a^2 + z_a^2)^{1/2} \\
&= (1^2 + 2^2 + 3^2)^{1/2} \\
&= (1 + 4 + 9)^{1/2} \\
&= 14^{1/2} \\
&= 3.7417
\end{aligned}$$

PROBLEM 6-8

Consider two vectors **a** and **b** in xyz-space, both of which lie in the xy-plane. The vectors are represented by the following ordered triples:

$$\mathbf{a} = (3,4,0)$$
$$\mathbf{b} = (0, -5, 0)$$

Find the ordered triple that represents the vector $\mathbf{a} \times \mathbf{b}$.

SOLUTION 6-8

Let's draw these two vectors as they appear in the xy-plane. See Fig. 6-12. In this drawing, imagine the positive z axis coming out of the page directly toward you, and the negative z axis pointing straight away from you on the other side of the page.

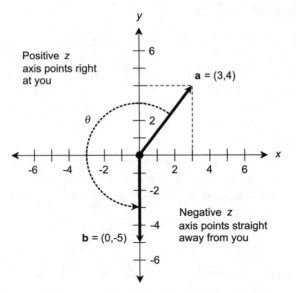

Fig. 6-12.   Illustration for Problem 6-8.

First, let's figure out the direction in which $\mathbf{a} \times \mathbf{b}$ points. The direction of the cross product of two vectors is always perpendicular to the plane containing the original vectors. Thus, $\mathbf{a} \times \mathbf{b}$ points along the z axis. The ordered triple must be in the form $(0,0,z)$, where z is some real number. We don't yet know what this number is, and we had better not jump to any conclusions. Is it positive? Negative? Zero? We must proceed further to find out.

Next, we calculate the lengths (magnitudes) of the two vectors $\mathbf{a}$ and $\mathbf{b}$. To find $|\mathbf{a}|$, we use the formula:

$$|\mathbf{a}| = (x_a^2 + y_a^2)^{1/2}$$
$$= (3^2 + 4^2)^{1/2}$$
$$= (9 + 16)^{1/2}$$
$$= 25^{1/2}$$
$$= 5$$

Similarly, for $|\mathbf{b}|$:

$$|\mathbf{b}| = (x_b^2 + y_b^2)^{1/2}$$
$$= [0^2 + (-5)^2]^{1/2}$$
$$= 25^{1/2}$$
$$= 5$$

Therefore, $|\mathbf{a}|\,|\mathbf{b}| = 5 \times 5 = 25$. In order to determine the magnitude of $\mathbf{a} \times \mathbf{b}$, we must multiply this by the sine of the angle θ between the two vectors, as expressed counterclockwise from $\mathbf{a}$ to $\mathbf{b}$.

To find the measure of θ, note that it is equal to 270° (three-quarters of a circle) minus the angle between the x axis and the vector $\mathbf{a}$. The angle between the x axis and vector $\mathbf{a}$ is the arctangent of 4/3, or approximately 53° as determined using a calculator. Therefore:

$$\theta = 270° - 53° = 217°$$
$$\sin\,\theta = \,\sin\,217° = -0.60 \text{ (approx.)}$$

This means that the magnitude of $\mathbf{a} \times \mathbf{b}$ is equal to approximately $25 \times (-0.60)$, or -15. The minus sign is significant. It means that the cross product vector points negatively along the z axis. Therefore, the z coordinate of $\mathbf{a} \times \mathbf{b}$ is equal to -15. We know that the x and y coordinates of $\mathbf{a} \times \mathbf{b}$ are both equal to 0 because $\mathbf{a} \times \mathbf{b}$ lies along the z axis. It follows that $\mathbf{a} \times \mathbf{b} = (0,0,-15)$.

Quiz

Refer to the text in this chapter if necessary. A good score is eight correct. Answers are in the back of the book.

1. The magnitude (or length) of the cross product of two vectors depends on
   (a) the angle between them
   (b) their magnitudes

(c) their magnitudes and the angle between them

(d) neither their magnitudes nor the angle between them

2. The dot product of two vectors depends on
   (a) the angle between them
   (b) their magnitudes
   (c) their magnitudes and the angle between them
   (d) neither their magnitudes nor the angle between them

3. In spherical coordinates, the position of a point is specified by
   (a) two angles and a distance
   (b) two distances and an angle
   (c) three distances
   (d) three angles

4. Suppose you see a balloon hovering in the sky over a calm ocean. You are told that it is at azimuth 30°, that it is 3500 meters above the ocean surface, and that the point directly underneath it is 5000 meters away from you. This information is an example of the position of the balloon expressed in a form of
   (a) Cartesian coordinates
   (b) cylindrical coordinates
   (c) spherical coordinates
   (d) celestial coordinates

5. Suppose vector **a**, represented in Cartesian three-space by $(3,-1,-5)$, is multiplied by a constant $k = 2$. What represents the product $k\mathbf{a}$?
   (a) The ordered triple $(6,-2,-10)$
   (b) The ordered triple $(-6,2,10)$
   (c) The scalar -6
   (d) Nothing; $k\mathbf{a}$ is not defined because a vector cannot be multiplied by a scalar

6. Suppose there are two vectors that correspond to the Cartesian ordered pairs $\mathbf{a} = (3,5)$ and $\mathbf{b} = (1,0)$. What is the dot product of these vectors?
   (a) $(3,0)$
   (b) $(4,5)$
   (c) $(0,0)$
   (d) None of the above

7. Two vectors point in opposite directions, and one has twice the magnitude of the other. Their cross product
   (a) points in the direction of the vector with the larger magnitude
   (b) points in the direction of the vector with the smaller magnitude

    (c) points in a direction perpendicular to the plane containing both
        vectors
    (d) has zero magnitude

8. A vector denotes a phenomenon that has
    (a) an abscissa and an ordinate
    (b) a radius and an angle
    (c) a magnitude and a direction
    (d) an azimuth and an elevation

9. In Cartesian three-space,
    (a) the axes are all mutually perpendicular
    (b) θ represents azimuth, ϕ represents elevation, and r represents
        radius
    (c) vectors are represented by an angle and a radius
    (d) coordinates are all defined by angles

10. Refer to Fig. 6-8 in the chapter text. What is the dot product of the
    vectors (θ,r) and (θ,kr)?
    (a) $-kr^2$
    (b) kr^2
    (c) k^2r^2
    (d) It is impossible to determine this without more information

Test: Part One

Do not refer to the text when taking this test. You may draw diagrams or use a calculator if necessary. A good score is at least 38 correct. Answers are in the back of the book. It's best to have a friend check your score the first time, so you won't memorize the answers if you want to take the test again.

1. With respect to the circular functions, an angle whose measure is equal to $-45°$ is the same as an angle whose measure is
   (a) $45°$
   (b) $135°$
   (c) $225°$
   (d) $315°$
   (e) undefined

2. Suppose there is a triangle whose sides are 5, 12, and 13 units, respectively. What is the sine of the angle opposite the side that measures 13 units, accurate to three decimal places?
   (a) 0.385
   (b) 0.417
   (c) 0.923
   (d) 1.000
   (e) It cannot be determined without more information

3. What is the arcsine of 3?
    (a) 30°
    (b) 60°
    (c) 90°
    (d) 180°
    (e) It is not defined

4. In Fig. Test 1-1, the solid curve represents the hyperbolic cosine function ($y = \cosh x$), and the dashed curve represents the inverse of the hyperbolic cosine function ($y = \text{arccosh } x$). From this, it appears that the domain of $f(x) = \text{arccosh } x$
    (a) includes all real numbers
    (b) includes all non-negative real numbers
    (c) includes all real numbers greater than 1
    (d) includes all real numbers greater than or equal to 1
    (e) does not include any real numbers

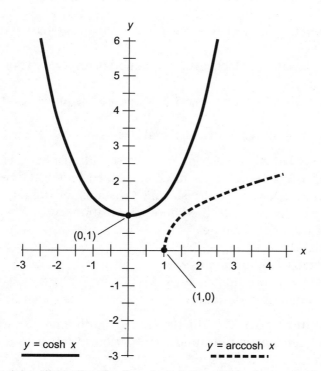

Fig. Test 1-1. Illustration for Questions 4, 5, and 6 in the test for Part One.

5. Based on the information shown in Fig. Test 1-1 and given in Question 4, it appears that the hyperbolic cosine of 0 is
   (a) equal to 1
   (b) equal to 0
   (c) non-negative
   (d) greater than or equal to 1
   (e) not defined

6. Based on the information shown in Fig. Test 1-1 and given in Question 4, it appears that the hyperbolic arccosine of 0 is
   (a) equal to 1
   (b) equal to 0
   (c) non-negative
   (d) greater than or equal to 1
   (e) not defined

7. Consider a system of Cartesian coordinates where x represents the independent variable and y represents the dependent variable. A function in this system is
   (a) a relation in which every x value corresponds to at least one y value
   (b) a relation in which every y value corresponds to at least one x value
   (c) a relation in which every x value corresponds to at most one y value
   (d) a relation in which every y value corresponds to at most one x value
   (e) not described by any of the above

8. Suppose a balloon, hovering high in the atmosphere, is located in a position with respect to an observer defined by the following: azimuth 45°, elevation 60°, radius (also called distance or range) 25 kilometers. This is an expression of the balloon's position in
   (a) Cartesian coordinates
   (b) polar coordinates
   (c) cylindrical coordinates
   (d) spherical coordinates
   (e) rectangular coordinates

9. In navigator's polar coordinates, the azimuth angle is measured
   (a) counterclockwise around the horizon, relative to a ray pointing east
   (b) clockwise around the horizon, relative to a ray pointing north

    (c) downward relative to a ray pointing toward the zenith

    (d) upward relative to a ray pointing toward the horizon

    (e) upward relative to a ray pointing toward the nadir

10. Suppose **v** is a vector in three-space. Let the magnitude of this vector be denoted v. What is **v** · **v** (the dot product of vector **v** with itself)?

    (a) $2v$

    (b) v^2

    (c) 0

    (d) 1

    (e) It is impossible to determine without more information

11. For which of the following angles is the value of the tangent function not defined?

    (a) 0 rad

    (b) $\pi/6$ rad

    (c) $\pi/4$ rad

    (d) $\pi/2$ rad

    (e) It is defined for all of the above values

12. In Fig. Test 1-2, which of the following ratios represents csc θ?

    (a) e/f

    (b) d/f

    (c) d/e

    (d) e/d

    (e) None of the above

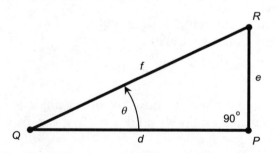

Fig. Test 1-2. Illustration for Questions 12, 13, and 14 in the test for Part One.

13. In Fig. Test 1-2, which of the following ratios represents cos θ?

    (a) e/f

    (b) d/f

    (c) d/e

(d) e/d

(e) None of the above

14. In Fig. Test 1-2, which of the following ratios represents cot θ?
    (a) e/f
    (b) d/f
    (c) d/e
    (d) e/d
    (e) None of the above

15. As long as the measure of an angle is not equal to any integer multiple of 180°, its sine is equal to the reciprocal of its
    (a) cosecant
    (b) cosine
    (c) secant
    (d) tangent
    (e) cotangent

16. Given that csch $x = 2/(e^x - e^{-x})$, what can be said about csch 0?
    (a) It is equal to zero
    (b) It is a positive real number less than 1
    (c) It is equal to 1
    (d) It is a positive real number greater than 1
    (e) It is undefined

17. If a vector is multiplied by 2, what happens to its orientation?
    (a) It does not change
    (b) It is shifted counterclockwise by 90°
    (c) It is shifted clockwise by 90°
    (d) It is shifted by 180°
    (e) It doubles

18. Suppose we are told that the measure of an angle θ lies somewhere between (but not including) 90° and 270°. We can be certain that the value of $\sin^2 \theta + \cos^2 \theta$ is
    (a) greater than 0 but less than 1
    (b) greater than −1 but less than 0
    (c) greater than −1 but less than 1
    (d) equal to 1
    (e) equal to 0

19. If a vector is multiplied by 2, what happens to its magnitude?
    (a) It does not change
    (b) It doubles
    (c) It is cut to $\frac{1}{2}$ its previous value
    (d) It quadruples
    (e) It is cut to $\frac{1}{4}$ its previous value

20. The range of the function $y = \sin x$ encompasses
    (a) all real numbers between but not including -1 and 1
    (b) all real numbers between but not including 0 and 1
    (c) all real numbers between and including -1 and 1
    (d) all real numbers between and including 0 and 1
    (e) all real numbers

21. Suppose a broadcast tower is constructed in a perfectly square field that measures 100 meters on each side. The tower is in the center of the field and is 50 meters high. It is guyed from the middle and from the top. The guy wires run to the corners of the field. At what angle, to the nearest degree and relative to the horizontal, do the top guy wires slant?
    (a) $35°$
    (b) $45°$
    (c) $55°$
    (d) $65°$
    (e) It cannot be determined without more information

22. In the graph of Fig. Test 1-3, suppose that $x_0 = 2.91$ and $y_0 = 3.58$. What is the value of r_0, rounded to two decimal places?
    (a) 3.25
    (b) 4.61
    (c) 6.49
    (d) 21.28
    (e) It cannot be determined without more information

23. In the graph of Fig. Test 1-3, suppose that $x_0 = 2.91$ and $y_0 = 3.58$. What is the value of θ_0, rounded to the nearest degree?
    (a) $36°$
    (b) $39°$
    (c) $51°$
    (d) $54°$
    (e) It cannot be determined without more information

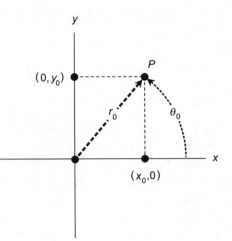

Fig. Test 1-3. Illustration for Questions 22, 23, and 24 in the test for Part One.

24. In the graph of Fig. Test 1-3, suppose that $\theta_0 = 45°$ and $r_0 = 5.35$. What is the Cartesian coordinate (x_0, y_0) of point P? Express both values rounded to two decimal places.
    (a) (3.78,3.78)
    (b) (5.35,5.35)
    (c) (2.31,2.31)
    (d) (2.68,2.68)
    (e) It cannot be determined without more information

25. What is the arcsine of 0.5?
    (a) 0°
    (b) 30°
    (c) 60°
    (d) 90°
    (e) It is not defined

26. Suppose you are standing on a flat, empty playing field and it is a sunny day. You measure the length of your shadow and discover that it is exactly twice your height. What is the angle of the sun above the horizon (if that angle can be determined) to the nearest degree?
    (a) 27°
    (b) 30°
    (c) 60°
    (d) 63°
    (e) It depends on how tall you are

27. Suppose you are standing on a flat, empty playing field and it is a sunny day. You measure the length of your shadow and discover that it is exactly 1 meter greater than your height. What is the angle of the sun above the horizon (if the angle can be determined) to the nearest degree?
    (a) 27°
    (b) 30°
    (c) 60°
    (d) 63°
    (e) It depends on how tall you are

28. Suppose a point is located at (x_0, y_0) in a Cartesian coordinate system. What is the radius, r, in mathematician's polar coordinates?
    (a) $r = x_0^2 + y_0^2$
    (b) $r = x_0^2 - y_0^2$
    (c) $r = (x_0^2 + y_0^2)/2$
    (d) $r = (x_0^2 - y_0^2)/2$
    (e) None of the above

29. Suppose there are two vectors, **a** and **b**, and that vector **a** points straight west while vector **b** points straight north. In what direction does vector **a** × **b** point?
    (a) Southeast
    (b) Northwest
    (c) Straight up
    (d) Straight down
    (e) This question has no answer, because **a** × **b** is a scalar, not a vector

30. A radian is the equivalent of
    (a) 2π angular degrees
    (b) $1/(2\pi)$ of the angle comprising a full circle
    (c) the circumference of a unit circle
    (d) $1/4$ of the angle comprising a full circle
    (e) $1/2$ of the circumference of a circle

31. Suppose we restrict the domain of $y = \sin x$ to allow only values of x between, but not including, −30° and 30°. What is the range of the resulting function?
    (a) $0 < y < 0.5$
    (b) $-0.5 < y < 0$
    (c) $-0.5 < y < 0.5$
    (d) The entire set of real numbers
    (e) It is undefined

32. Given that arccsch $x = \ln [x^{-1} + (x^{-2} + 1)^{1/2}]$, what can be said about arccsch 0?
    (a) It is equal to zero
    (b) It is a positive real number less than 1
    (c) It is equal to 1
    (d) It is a positive real number greater than 1
    (e) It is undefined

33. In mathematician's polar coordinates, an angle of $-90°$ is equivalent to an angle of
    (a) $\pi/4$ rad
    (b) $\pi/2$ rad
    (c) $3\pi/4$ rad
    (d) $5\pi/4$ rad
    (e) None of the above

34. Suppose an antenna tower is 250 meters high and stands in a perfectly flat field. The highest set of guy wires comes down from the top of the tower at a 45° angle relative to the tower itself. How long is each of these guy wires? Express your answer (if an answer exists) to the nearest meter.
    (a) 250 meters
    (b) 354 meters
    (c) 375 meters
    (d) 400 meters
    (e) It is impossible to tell without more information

35. What does the graph of the equation $r = -\theta/(20\pi)$ look like in mathematician's polar coordinates, when θ is expressed in radians?
    (a) A large circle
    (b) A 20-leafed clover
    (c) A large cardioid
    (d) A tightly wound spiral
    (e) Nothing, because the radius must always be negative, and such a condition is not defined

36. What is the vector sum $\mathbf{a} + \mathbf{b}$ in Fig. Test 1-4?
    (a) (7.4,0.4)
    (b) (3.9,6.1)
    (c) (1.7,6.1)
    (d) (6.1,1.7)
    (e) (0.4,7.4)

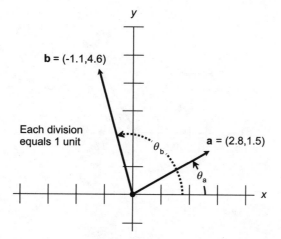

b = (-1.1,4.6)

Each division
equals 1 unit

θ_b

a = (2.8,1.5)

θ_a

y

x

Fig. Test 1-4. Illustration for Questions 36, 37, and 38 in the test for Part One.

37. What is the angle θ_a in Fig. Test 1-4 to the nearest degree?
    (a) 28°
    (b) 31°
    (c) 32°
    (d) 35°
    (e) It cannot be determined without more information

38. What is the angle θ_b in Fig. Test 1-4?
    (a) arctan $[4.6/(-1.1)]$
    (b) π + arctan $[(4.6/(-1.1)]$
    (c) arctan $(-1.1/4.6)$
    (d) π + arctan $(-1.1/4.6)$
    (e) It cannot be determined without more information

39. Suppose, in a set of mathematician's polar coordinates, an object has a
    radius coordinate $r = 7$ units. What is the angle coordinate θ of the
    object in radians?
    (a) $\theta = \pi/2$
    (b) $\theta = \pi$
    (c) $\theta = 3\pi/2$
    (d) $\theta = 2\pi$
    (e) It cannot be determined without more information

40. In Fig. Test 1-5, the x value of point Q is equal to
    (a) $\sin \theta$
    (b) $\cos \theta$
    (c) $\tan \theta$

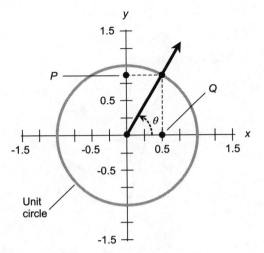

Fig. Test 1-5. Illustration for Questions 40, 41, and 42 in the test for Part One.

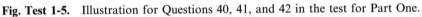

(d) arctan θ
(e) arccos θ

41. In Fig. Test 1-5, the y value of point P is equal to
    (a) sin θ
    (b) cos θ
    (c) tan θ
    (d) arctan θ
    (e) arccos θ

42. In Fig. Test 1-5, let q represent the x value of point Q, and let p represent the y value of point P. Which of the following statements is false?
    (a) arcsin $p = \theta$
    (b) arccos $q = \theta$
    (c) $q^2 + p^2 = 1$
    (d) $p/q = \tan \theta$
    (e) $1/p = \cot \theta$

43. Suppose an object is located 30 kilometers south and 30 kilometers west of the origin in a set of navigator's polar coordinates. What is the azimuth of this object?
    (a) 45°
    (b) 135°
    (c) 225°
    (d) 315°

(e) It is not defined

44. In Fig. Test 1-6, it is apparent that if the curve represents a trigono-
    metric relation, then
    (a) the x axis is graduated in radians
    (b) the y axis is graduated in radians
    (c) both the x and the y axes are graduated in radians

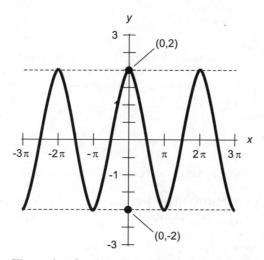

Fig. Test 1-6. Illustration for Questions 44, 45, and 46 in the test for Part One.

    (d) neither the x axis nor the y axis is graduated in radians
    (e) the system of coordinates is sinusoidal

45. By inspecting Fig. Test 1-6, we can conclude that
    (a) the wave-shaped curve represents a relation between x and y, but
        not a function of x
    (b) the wave-shaped curve represents a function of x, but not a rela-
        tion between x and y
    (c) the wave-shaped curve represents a function of x
    (d) the wave-shaped curve represents a function of y
    (e) None of the above

46. Suppose we are told that the curve in Fig. Test 1-6 has a sinusoidal
    shape. Also suppose that the maximum y value attained by the curve is
    2, and the minimum y value is −2 (as shown by the dashed line and the
    labeled points). From this it is apparent that
    (a) the curve represents the graph of $y = (\sin x)/2$
    (b) the curve represents the graph of $y = (\cos x)/2$

(c) the curve represents the graph of $y = 2 \sin x$

(d) the curve represents the graph of $y = 2 \cos x$

(e) the curve represents the graph of $y = \sin 2x$

47. What is the value of cosh [arccosh $(k + 1)$], where k is a positive real number?

  (a) $k + 1$

  (b) $k - 1$

  (c) $e^k + e$

  (d) $\ln (k - 1)$

  (e) It cannot be determined without more information

48. The sum of the measures of the interior angles of a triangle is

  (a) $\pi/2$ rad

  (b) π rad

  (c) $3\pi/2$ rad

  (d) 2π rad

  (e) dependent on the relative lengths of the sides

49. The hyperbolic functions arise from a curve that has a certain equation in the rectangular xy-plane. What is that equation?

  (a) $y = x^2 + 2x + 1$

  (b) $y = x^2$

  (c) $x = y^2$

  (d) $x^2 + y^2 = 1$

  (e) $x^2 - y^2 = 1$

50. One second of arc is equal to

  (a) $\pi/60$ radians

  (b) $\pi/3600$ radians

  (c) $1/60$ of an angular degree

  (d) $1/3600$ of an angular degree

  (e) a meaningless expression

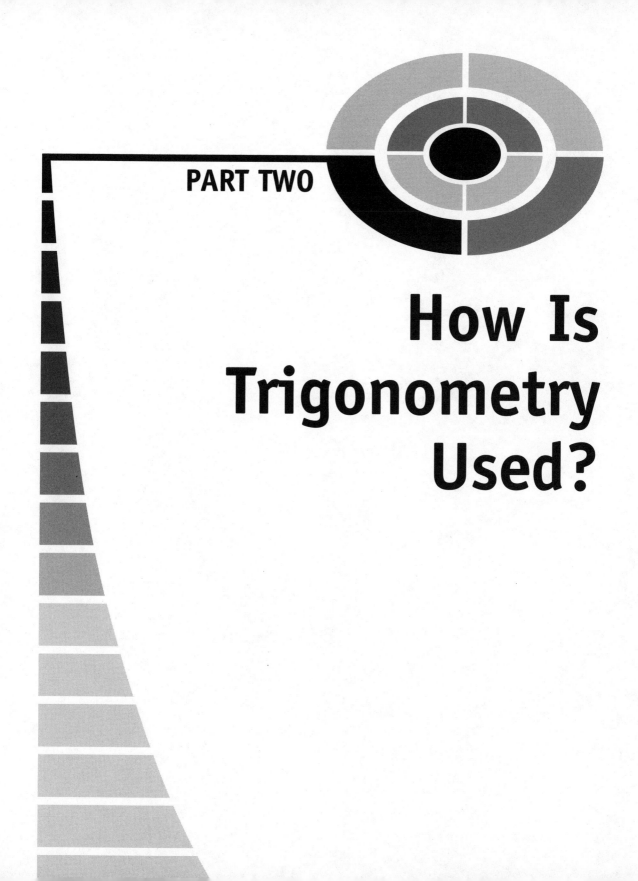

PART TWO

How Is Trigonometry Used?

Scientific Notation

Scientists, engineers, and other technical people use *scientific notation* to express the extreme quantitative values they encounter. In real-world applications, trigonometry often involves vast distances and tiny angles that don't lend themselves very well to expression as ordinary decimal numbers.

Some of the problems so far in this book have involved rounding answers off to a certain number of decimal places. In scientific and engineering work, it is the number of *significant figures*, more than the number of decimal places, that matters. Decimal places and significant figures sometimes mean the same thing, but not always.

Subscripts and Superscripts

Subscripts are used to modify the meanings of units, constants, and variables. A subscript is placed to the right of the main character (without spacing) and is set below the base line.

Superscripts almost always represent *exponents* (the raising of a base quantity to a power). Italicized, lowercase English letters from the second half of the alphabet (n through z) denote variable exponents. A superscript is placed to the right of the main character (without spacing) and is set above the base line.

EXAMPLES OF SUBSCRIPTS

Numeric subscripts are never italicized, but alphabetic subscripts are if they represent variables. Here are three examples of subscripted quantities:

- θ_0 – read "theta sub nought"; stands for a specific angle
- R_{out} – read "R sub out"; stands for output resistance in an electronic circuit
- y_n – read "y sub n"; represents a variable with a variable subscript

Ordinary numbers are rarely, if ever, modified with subscripts. You are not likely to see expressions like this:

$$3_5$$
$$-9.7755_\pi$$
$$16_x$$

Constants and variables can come in many "flavors." Some physical constants are assigned subscripts by convention. An example is m_e, representing the mass of an electron at rest. (The "e" in this case is not italicized because it stands for the word "electron," not for a variable or the natural logarithm base.)

Sometimes subscripts are used for convenience. Points in three-dimensional space are sometimes denoted using ordered triples such as (x_1, x_2, x_3) rather than (x, y, z). This subscripting scheme becomes especially convenient if you're talking about points in a higher-dimensional space, for example $(x_1, x_2, x_3, ..., x_{11})$ in Cartesian 11-dimensional (11D) space.

EXAMPLES OF SUPERSCRIPTS

Numeric superscripts are never italicized, but alphabetic superscripts usually are. Examples of superscripted quantities are:

- 2^3 – read "two cubed"; represents $2 \times 2 \times 2$
- $\sin^2 \theta$ – read "the square of the sine of theta"; represents a quantity multiplied by itself
- $\sin^{-1} \theta$ – read "the inverse sine of theta"; alternative expression for arcsin θ

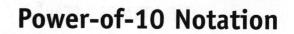

Power-of-10 Notation

Scientists and engineers like to express extreme numerical values using an exponential technique known as *power-of-10 notation*. This is usually what is meant when they talk about scientific notation.

STANDARD FORM

A numeral in *standard power-of-10 notation* is written as follows:

$$m.n \times 10^z$$

where the dot (.) is a period, written on the base line (not a raised dot indicating multiplication), and is called the *radix point* or *decimal point*. The value m (to the left of the radix point) is a positive integer from the set $\{1, 2, 3, 4, 5, 6, 7, 8, 9\}$. The value n (to the right of the radix point) is a non-negative integer. The value z, which is the power of 10, can be any integer: positive, negative, or zero. Here are some examples of numbers written in standard scientific notation:

$$2.56 \times 10^6$$
$$8.0773 \times 10^{-18}$$
$$1.000 \times 10^0$$

ALTERNATIVE FORM

In certain countries, and in many books and papers written before the middle of the 20th century, a slight variation on the above theme is used. The *alternative power-of-10 notation* requires that m be 0 rather than 1, 2, 3, 4, 5, 6, 7, 8, or 9. When the above quantities are expressed this way, they appear as decimal fractions larger than 0 but less than 1, and the value of the exponent is increased by 1 compared with the standard form:

$$0.256 \times 10^7$$
$$0.80773 \times 10^{-17}$$
$$0.1000 \times 10^1$$

These are the same three numerical values as the previous three; the only difference is the way they're expressed. It's like saying you're driving down a road at 50,000 meters per hour rather than at 50 kilometers per hour.

THE "TIMES SIGN"

The multiplication sign in a power-of-10 expression can be denoted in various ways. Most scientists in America use the cross symbol ($\times$), as in the examples shown above. But a small dot raised above the base line ($\cdot$) is sometimes used to represent multiplication in power-of-10 notation. When written that way, the above numbers look like this in the standard form:

$$2.56 \cdot 10^6$$
$$8.0773 \cdot 10^{-18}$$
$$1.000 \cdot 10^0$$

This small dot should not be confused with a radix point, as in the expression

$$m.n \cdot 10^z$$

in which the dot between m and n is a radix point and lies along the base line, while the dot between n and 10^z is a multiplication symbol and lies above the base line. The small dot is preferred when multiplication is required to express the dimensions of a physical unit. An example is the kilogram-meter per second squared, which is symbolized $kg \cdot m/s^2$ or $kg \cdot m \cdot s^{-2}$.

When using an old-fashioned typewriter, or in word processors that lack a good repertoire of symbols, the lowercase, non-italicized letter x can be used to indicate multiplication. But this can cause confusion, because it's easy to mistake this letter x for a variable. So in general, it's a bad idea to use the letter x as a "times sign." An alternative in this situation is to use an asterisk (*). This is why you will occasionally see numbers written like this:

$$2.56*10^6$$
$$8.0773*10^{-18}$$
$$1.000*10^0$$

PLAIN-TEXT EXPONENTS

Once in a while, you will have to express numbers in power-of-10 notation using plain, unformatted text. This is the case, for example, when transmitting information within the body of an e-mail message. Some calculators and computers use this system. An uppercase or lowercase letter E indicates that the quantity immediately following is a power of 10. The power-of-10 designator always includes a sign (plus or minus) unless it is zero. In this format, the above quantities are written like this:

$$2.56E + 6$$
$$8.0773E - 18$$
$$1.000E0$$

or like this:

$$2.56e + 6$$
$$8.0773e - 18$$
$$1.000e0$$

Another alternative is to use an asterisk to indicate multiplication, and the symbol ^ to indicate a superscript, so the expressions look like this:

$$2.56*10^6$$
$$8.0773*10^\wedge - 18$$
$$1.000*10^0$$

In all of the above examples, the numerical values, written out in fully expanded decimal form, look like this:

$$2,560,000$$
$$0.0000000000000000080773$$
$$1.000$$

ORDERS OF MAGNITUDE

As you can see, power-of-10 notation makes it possible to easily write down numbers that denote huge or tiny quantities. Consider the following:

$$2.55 \times 10^{45,589}$$
$$-9.8988 \times 10^{-7,654,321}$$

Imagine the task of writing either of these numbers out in ordinary decimal form! In the first case, you'd have to write the numerals 255, and then follow them with a string of 45,587 zeros. In the second case, you'd have to write a minus sign, then a numeral zero, then a radix point, then a string of 7,654,320 zeros, then the numerals 9, 8, 9, 8, and 8.

Now consider these two numbers:

$$2.55 \times 10^{45,592}$$
$$-9.8988 \times 10^{-7,654,318}$$

These resemble the first two, don't they? But they are vastly different. Both of these new numbers are a thousand times larger than the original two. You can tell by looking at the exponents. Both exponents are larger by three. The number 45,592 is three more than 45,589, and the number −7,754,318 is three larger than −7,754,321. (Numbers grow larger in the mathematical sense as they become more positive or less negative.) The second pair of numbers in the scientific-notation example above are both three *orders of magnitude* larger than the first pair of numbers. They look almost the same here, but they are as different as a meter compared to a kilometer, or a gram compared to a kilogram.

The order-of-magnitude concept makes it possible to construct number lines, charts, and graphs with scales that cover huge spans of values. Three examples are shown in Fig. 7-1. Drawing A shows a number line spanning three orders of magnitude, from 1 to 1000. Illustration B shows a number line spanning 10 orders of magnitude, from 10^{-3} to 10^7. Illustration C shows a graph whose horizontal scale spans 10 orders of magnitude, from 10^{-3} to 10^7, and whose vertical scale extends from 0 to 10.

If you're astute, you'll notice that while the 0-to-10 linear scale is the easiest to directly envision, it covers infinitely many orders of magnitude! This is because, no matter how many times you cut a nonzero number to 1/10 its original size, you can never reach zero.

PREFIX MULTIPLIERS

Special verbal prefixes, known as *prefix multipliers*, are commonly used by physicists and engineers to express orders of magnitude in power-of-10 notation. Table 7-1 shows the prefix multipliers used for factors ranging from 10^{-24} to 10^{24}. You've come across some of these. Your computer has a processor with a frequency of a certain number of gigahertz (multiples of 10^9 cycles per second). The moon is about 4×10^5 kilometers (multiples of 10^3 meters) away from the earth.

PROBLEM 7-1
Express the angle 0° 0′ 5″ in scientific notation. Then determine the difference, in orders of magnitude, between angles 180° and 0° 0′ 5″.

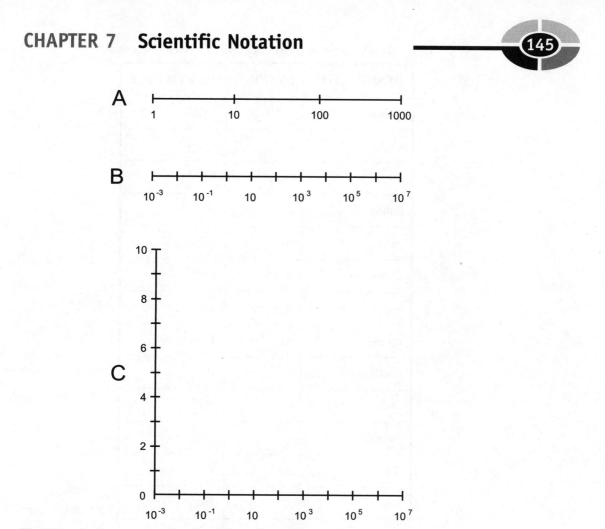

Fig. 7-1.   At A, a number line spanning three orders of magnitude. At B, a number line spanning 10 orders of magnitude. At C, a coordinate system whose horizontal scale spans 10 orders of magnitude, and whose vertical scale extends from 0 to 10.

SOLUTION 7-1

Recall that the apostrophe or accent symbol ($'$) means one minute of arc, or 1/60 of an angular degree, and the double apostrophe or accent ($''$) means one second of arc, which is 1/60 of an arc minute or 1/3600 of an angular degree. Thus:

$$0° \; 0' \; 5'' = (5/3600)°$$
$$= 0.00138888 \ldots °$$
$$= (1.38888 \ldots \times 10^{-3})°$$

Table 7-1. Power-of-10 prefix multipliers and their abbreviations

DESIGNATOR	SYMBOL	MULTIPLIER
yocto-	y	10^{-24}
zepto-	z	10^{-21}
atto-	a	10^{-18}
femto-	f	10^{-15}
pico-	p	10^{-12}
nano-	n	10^{-9}
micro-	μ or mm	10^{-6}
milli-	m	10^{-3}
centi-	c	10^{-2}
deci-	d	10^{-1}
(none)	–	10^{0}
deka-	da or D	10^{1}
hecto-	h	10^{2}
kilo-	K or k	10^{3}
mega-	M	10^{6}
giga-	G	10^{9}
tera-	T	10^{12}
peta-	P	10^{15}
exa-	E	10^{18}
zetta-	Z	10^{21}
yotta-	Y	10^{24}

Divide $180°$ by $(5/3600)°$ to determine how many times as large the big angle is compared to the tiny one:

$$180°/(5/3600)° = (180 \times 3600)/5$$
$$= 129,600$$
$$= 1.296 \times 10^5$$

This means that $180°$ is approximately five orders of magnitude larger than $0° \ 0' \ 5''$.

PROBLEM 7-2
What is the sine of the angle $0° \ 0' \ 5''$?

SOLUTION 7-2
This requires a calculator that can display many digits, and ideally, one that can work in scientific notation and that defaults to this notation when numbers become extreme. Here is what happens when the calculator in Microsoft Windows™ 98 is used to solve this problem. First, convert the angle to a decimal number of degrees:

$$0° \ 0' \ 5'' = (5/3600)°$$
$$= 0.00138888 \ldots °$$

Then click on the "sin" button, making sure that the "Dec" (for "decimal") and "Degrees" options are checked:

$$\sin 0.00138888 \ldots °$$
$$= 2.42 \ldots e{-}5$$
$$= 2.42 \times 10^{-5} \text{ (rounded to two decimal places)}$$

Rules for Use

In printed literature, power-of-10 notation is generally used only when the power of 10 is large or small. If the exponent is between -2 and 2 inclusive, numbers are written out in plain decimal form as a rule. If the exponent is -3 or 3, numbers are sometimes written out, and are sometimes written in power-of-10 notation. If the exponent is -4 or smaller, or if it is 4 or larger, values are expressed in power-of-10 notation as a rule.

Some calculators, when set for power-of-10 notation, display all numbers that way. This can be confusing, especially when the power of 10 is zero and the calculator is set to display a lot of digits. Most people understand the expression 8.407 more easily than 8.407000000e+0, for example, even though they represent the same number.

With that in mind, let's see how power-of-10 notation works when we want to do simple arithmetic using extreme numbers.

ADDITION

Addition of numbers is best done by writing numbers out in ordinary decimal form if at all possible. Thus, for example:

$$(3.045 \times 10^5) + (6.853 \times 10^6)$$
$$= 304,500 + 6,853,000$$
$$= 7,157,500$$
$$= 7.1575 \times 10^6$$

$$(3.045 \times 10^{-4}) + (6.853 \times 10^{-7})$$
$$= 0.0003045 + 0.0000006853$$
$$= 0.0003051853$$
$$= 3.051853 \times 10^{-4}$$

$$(3.045 \times 10^5) + (6.853 \times 10^{-7})$$
$$= 304,500 + 0.0000006853$$
$$= 304,500.0000006853$$
$$= 3.045000000006853 \times 10^5$$

SUBTRACTION

Subtraction follows the same basic rules as addition:

$$(3.045 \times 10^5) - (6.853 \times 10^6)$$
$$= 304,500 - 6,853,000$$
$$= -6,548,500$$
$$= -6.548500 \times 10^6$$

$$(3.045 \times 10^{-4}) - (6.853 \times 10^{-7})$$
$$= 0.0003045 - 0.0000006853$$
$$= 0.0003038147$$
$$= 3.038147 \times 10^{-4}$$

$$(3.045 \times 10^{5}) - (6.853 \times 10^{-7})$$
$$= 304{,}500 - 0.0000006853$$
$$= 304{,}499.9999993147$$
$$= 3.044999999993147 \times 10^{5}$$

If the absolute values of two numbers differ by many orders of magnitude, the one with the smaller absolute value (that is, the one closer to zero) can vanish into insignificance, and for practical purposes, can be ignored. We'll look at that phenomenon later in this chapter.

MULTIPLICATION

When numbers are multiplied in power-of-10 notation, the decimal numbers (to the left of the multiplication symbol) are multiplied by each other. Then the powers of 10 are added. Finally, the product is reduced to standard form. Here are three examples, using the same three number pairs as before:

$$(3.045 \times 10^{5}) \times (6.853 \times 10^{6})$$
$$= (3.045 \times 6.853) \times (10^{5} \times 10^{6})$$
$$= 20.867385 \times 10^{(5+6)}$$
$$= 20.867385 \times 10^{11}$$
$$= 2.0867385 \times 10^{12}$$

$$(3.045 \times 10^{-4}) \times (6.853 \times 10^{-7})$$
$$= (3.045 \times 6.853) \times (10^{-4} \times 10^{-7})$$
$$= 20.867385 \times 10^{(-4-7)}$$
$$= 20.867385 \times 10^{-11}$$
$$= 2.0867385 \times 10^{-10}$$

$$(3.045 \times 10^5) \times (6.853 \times 10^{-7})$$
$$= (3.045 \times 6.853) \times (10^5 \times 10^{-7})$$
$$= 20.867385 \times 10^{(5-7)}$$
$$= 20.867385 \times 10^{-2}$$
$$= 2.0867385 \times 10^{-1}$$
$$= 0.20867385$$

This last number is written out in plain decimal form because the exponent is between −2 and 2 inclusive.

DIVISION

When numbers are divided in power-of-10 notation, the decimal numbers (to the left of the multiplication symbol) are divided by each other. Then the powers of 10 are subtracted. Finally, the quotient is reduced to standard form. Let's go another round with the same three number pairs we've been using:

$$(3.045 \times 10^5)/(6.853 \times 10^6)$$
$$= (3.045/6.853) \times (10^5/10^6)$$
$$\approx 0.444331 \times 10^{(5-6)}$$
$$= 0.444331 \times 10^{-1}$$
$$= 0.0444331$$

$$(3.045 \times 10^{-4})/(6.853 \times 10^{-7})$$
$$= (3.045/6.853) \times (10^{-4}/10^{-7})$$
$$\approx 0.444331 \times 10^{[-4-(-7)]}$$
$$= 0.444331 \times 10^3$$
$$= 4.44331 \times 10^2$$
$$= 444.331$$

$$(3.045 \times 10^5)/(6.853 \times 10^{-7})$$
$$= (3.045/6.853) \times (10^5/10^{-7})$$
$$\approx 0.444331 \times 10^{[5-(-7)]}$$
$$= 0.444331 \times 10^{12}$$
$$= 4.44331 \times 10^{11}$$

Note the "approximately equal to" signs ($\approx$) in the above equations. The quotients here don't divide out neatly to produce resultants with reasonable numbers of digits. To this, you might naturally ask, "How many digits is reasonable?" The answer lies in the method scientists use to determine significant figures. An explanation is coming up soon.

EXPONENTIATION

When a number is raised to a power in scientific notation, both the coefficient and the power of 10 must be raised to that power, and the result multiplied. Consider this:

$$(4.33 \times 10^5)^3$$
$$= 4.33^3 \times (10^5)^3$$
$$= 81.182737 \times 10^{(5 \times 3)}$$
$$= 81.182737 \times 10^{15}$$
$$= 8.1182737 \times 10^{16}$$

Let's consider another example, in which an exponent is negative:

$$(5.27 \times 10^{-4})^2$$
$$= 5.27^2 \times (10^{-4})^2$$
$$= 27.7729 \times 10^{(-4 \times 2)}$$
$$= 27.7729 \times 10^{-8}$$
$$= 2.77729 \times 10^{-7}$$

TAKING ROOTS

To find the root of a number in power-of-10 notation, the easiest thing to do is to consider that the root is a fractional exponent. The square root is the same thing as the $\frac{1}{2}$ power, the cube root is the same thing as the $\frac{1}{3}$ power, and so on. Then you can multiply things out in the same way as you would with whole-number powers. Here is an example:

$$(5.27 \times 10^{-4})^{1/2}$$

$$= (5.27)^{1/2} \times (10^{-4})^{1/2}$$

$$\approx 2.2956 \times 10^{[-4 \times (1/2)]}$$

$$= 2.2956 \times 10^{-2}$$

$$= 0.02956$$

Note, again, the "squiggly equals" sign. The square root of 5.27 is an irrational number, and the best we can do is to approximate its decimal expansion.

Approximation, Error, and Precedence

In trigonometry, the numbers we work with are rarely exact values. We must almost always settle for an approximation. There are two ways of doing this: *truncation* (easy but not very accurate) and *rounding* (a little trickier, but more accurate).

TRUNCATION

The process of truncation deletes all the numerals to the right of a certain point in the decimal part of an expression. Some electronic calculators use this process to fit numbers within their displays. For example, the number 3.830175692803 can be shortened in steps as follows:

3.830175692803

3.83017569280

3.8301756928

3.830175692

3.83017569

3.8301756

3.830175

3.83017

3.83

3.8

3

ROUNDING

Rounding is the preferred method of rendering numbers in shortened form. In this process, when a given digit (call it r) is deleted at the right-hand extreme of an expression, the digit q to its left (which becomes the new r after the old r is deleted) is not changed if $0 \leq r \leq 4$. If $5 \leq r \leq 9$, then q is increased by 1 ("rounded up"). The better electronic calculators use rounding rather than truncation. If rounding is used, the number 3.830175692803 can be shortened in steps as follows:

$$3.830175692803$$
$$3.83017569280$$
$$3.8301756928$$
$$3.830175693$$
$$3.83017569$$
$$3.8301757$$
$$3.830176$$
$$3.83018$$
$$3.8302$$
$$3.830$$
$$3.83$$
$$3.8$$
$$4$$

ERROR

When physical quantities are measured, exactness is impossible. Errors occur because of imperfections in the instruments, and in some cases because of human observational shortcomings or outright mistakes.

Suppose x_a represents the actual value of a quantity to be measured. Let x_m represent the measured value of that quantity, in the same units as x_a. Then the *absolute error*, D_a (in the same units as x_a), is given by:

$$D_a = x_m - x_a$$

The *proportional error*, D_p, is equal to the absolute error divided by the actual value of the quantity:

$$D_p = (x_m - x_a)/x_a$$

The *percentage error*, $D_\%$, is equal to 100 times the proportional error expressed as a ratio:

$$D_\% = 100(x_m - x_a)/x_a$$

Error values and percentages are positive if $x_m > x_a$, and negative if $x_m < x_a$. That means that if the measured value is too large, the error is positive, and if the measured value is too small, the error is negative. Sometimes the possible error or uncertainty in a situation is expressed in terms of "plus or minus" a certain number of units or percent. This is indicated by a *plus-or-minus sign* ($\pm$).

Note the denominators above that contain x_a, the actual value of the quantity under scrutiny, a quantity we don't exactly know because our measurement is imperfect! How can we calculate an error based on formulas containing a quantity subject to the very error in question? The common practice is to derive a theoretical or "ideal" value of x_a from scientific equations, and then compare the observed value to the theoretical value. Sometimes the observed value is obtained by taking numerous measurements, each with its own value x_{m1}, x_{m2}, x_{m3}, and so on, and then averaging them all.

PRECEDENCE

Mathematicians, scientists, and engineers have all agreed on a certain order in which operations should be performed when they appear together in an expression. This prevents confusion and ambiguity. When various operations such as addition, subtraction, multiplication, division, and exponentiation appear in an expression, and if you need to simplify that expression, perform the operations in the following sequence:

- Simplify all expressions within parentheses, brackets, and braces from the inside out
- Perform all exponential operations, proceeding from left to right
- Perform all products and quotients, proceeding from left to right
- Perform all sums and differences, proceeding from left to right

Here are two examples of expressions simplified according to the above rules of precedence. Note that the order of the numerals and operations is the same in each case, but the groupings differ.

$$[(2+3)(-3-1)^2]^2$$
$$[5 \times (-4)^2]^2$$
$$(5 \times 16)^2$$
$$80^2$$
$$6400$$

$$\{[2+3 \times (-3)-1]^2\}^2$$
$$[(2+(-9)-1)^2]^2$$
$$(-8^2)^2$$
$$64^2$$
$$4096$$

Suppose you're given a complicated expression and there are no parentheses in it? This does not have to be ambiguous, as long as the above-mentioned rules are followed. Consider this example:

$$z = -3x^3 + 4x^2y - 12xy^2 - 5y^3$$

If this were written with parentheses, brackets, and braces to emphasize the rules of precedence, it would look like this:

$$z = [-3(x^3)] + \{4[(x^2)y]\} - \{12[x(y^2)]\} - [5(y^3)]$$

Because we have agreed on the rules of precedence, we can do without the parentheses, brackets, and braces.

There is a certain elegance in minimizing the number of parentheses, brackets, and braces in mathematical expressions. But extra ones do no harm if they're placed correctly. You're better off to use a couple of unnecessary markings than to risk having someone interpret an expression the wrong way.

PROBLEM 7-3
Suppose you are given two vectors in mathematician's polar coordinates, as follows:

$$\mathbf{a} = (\theta_a, r_a) = [0°, (3.566 \times 10^{13})]$$
$$\mathbf{b} = (\theta_b, r_b) = [54°, (1.234 \times 10^7)]$$

Find the dot product $\mathbf{a} \cdot \mathbf{b}$, accurate to three decimal places by rounding.

SOLUTION 7-3

The dot product of two vectors in mathematician's polar coordinates is found by multiplying their lengths, and then multiplying this product by the cosine of the angle between the vectors. First, let's multiply their lengths, which happen to be r_a and r_b:

$$|\mathbf{a}||\mathbf{b}| = r_a r_b = (3.566 \times 10^{13}) \times (1.234 \times 10^7)$$
$$= 3.566 \times 1.234 \times 10^{(13+7)}$$
$$= 4.400444 \times 10^{20}$$

The angle θ between the vectors is $\theta_a - \theta_b = 54° - 0° = 54°$, and its cosine, found using a calculator, is 0.587785 (rounded to six decimal places). Thus, multiplying out and rounding to three decimal places, we get this:

$$|\mathbf{a}||\mathbf{b}| \cos \theta \approx 4.400444 \times 10^{20} \times 0.587785$$
$$\approx (4.400444 \times 0.587785) \times 10^{20}$$
$$\approx 2.587 \times 10^{20}$$

PROBLEM 7-4

Suppose you are given the same two vectors in mathematician's polar coordinates as those in the previous problem:

$$\mathbf{a} = (\theta_a, r_a) = [0°, (3.566 \times 10^{13})]$$
$$\mathbf{b} = (\theta_b, r_b) = [54°, (1.234 \times 10^7)]$$

Find the cross product $\mathbf{a} \times \mathbf{b}$, accurate to three decimal places by truncation.

SOLUTION 7-4

First, we should clarify the direction in which the cross product vector points. Remember the right-hand rule. Because the angle between vectors $\mathbf{a}$ and $\mathbf{b}$ is between 0° and 180° (non-inclusive), the vector $\mathbf{a} \times \mathbf{b}$ points up out of the page. Another way to envision this is that if the axis representing $\theta = 0$ in the mathematician's polar plane points due east and the angles are measured counterclockwise as viewed from above, then the vector $\mathbf{a} \times \mathbf{b}$ points toward the zenith.

The magnitude of $\mathbf{a} \times \mathbf{b}$ is found by multiplying their lengths, and then multiplying this product by the sine of the angle between the vectors. First, let's multiply their lengths, r_a and r_b:

$$|\mathbf{a}||\mathbf{b}| = r_a r_b = (3.566 \times 10^{13}) \times (1.234 \times 10^7)$$
$$= 3.566 \times 1.234 \times 10^{(13+7)}$$
$$= 4.400444 \times 10^{20}$$

The angle θ between the vectors is $54°$, just as it is in Problem 7-3. Its sine, found using a calculator, is 0.809016 (truncated to six decimal places). Multiplying out and truncating to three decimal places, we get this:

$$|\mathbf{a}||\mathbf{b}| \sin \theta \approx 4.400444 \times 10^{20} \times 0.809016$$
$$\approx (4.400444 \times 0.809016) \times 10^{20}$$
$$\approx 3.560 \times 10^{20}$$

Significant Figures

When multiplication or division is done using power-of-10 notation, the number of significant figures (also called *significant digits*) in the result cannot legitimately be greater than the number of significant figures in the least-exact expression.

Consider the two numbers $x = 2.453 \times 10^4$ and $y = 7.2 \times 10^7$. The following is a valid statement in pure arithmetic:

$$xy = 2.453 \times 10^4 \times 7.2 \times 10^7$$
$$= 2.453 \times 7.2 \times 10^{(4+7)}$$
$$= 17.6616 \times 10^{11}$$
$$= 1.76616 \times 10^{12}$$

But if x and y represent measured quantities, as they would in experimental science or engineering, the above statement needs qualification. We must pay close attention to how much accuracy we claim.

HOW ACCURATE ARE WE?

When you see a product or quotient containing a bunch of quantities in scientific notation, count the number of individual numerals (digits) in the decimal portions of each quantity. Then identify the quantity with the smallest number of digits, and count the number of individual numerals in it. That's the number of significant figures you can claim in the final answer or solution.

In the above example, there are four single digits in the decimal part of x, and two single digits in the decimal part of y. So we must round off the answer, which appears to contain six significant figures, to two. (It is important to use rounding, and not truncation!) We should conclude that:

$$xy = 2.453 \times 10^4 \times 7.2 \times 10^7$$
$$\approx 1.8 \times 10^{12}$$

NO MORE SQUIGGLIES

In science and engineering, approximation is the rule, not the exception. If you want to be rigorous, therefore, you must use squiggly equals signs whenever you round off any quantity, or whenever you make any observation. But writing these squigglies can get tiresome. Most scientists and engineers are content to use ordinary equals signs when it is understood there is an approximation or error involved in the expression of a quantity. From now on, let us do the same. No more squigglies!

Suppose we want to find the quotient x/y in the above situation, instead of the product xy. Proceed as follows:

$$x/y = (2.453 \times 10^4)/(7.2 \times 10^7)$$
$$= (2.453/7.2) \times 10^{(4-7)}$$
$$0.3406944444\ldots \times 10^{-3}$$
$$3.406944444\ldots \times 10^{-4}$$
$$= 3.4 \times 10^{-4}$$

WHAT ABOUT ZEROS?

Sometimes, when you make a calculation, you'll get an answer that lands on a neat, seemingly whole-number value. Consider $x = 1.41421$ and $y = 1.41422$. Both of these have six significant figures. The product, taking significant figures into account, is:

$$xy = 1.41421 \times 1.4142$$
$$= 2.0000040662$$
$$= 2.00000$$

This looks like it's exactly equal to 2. In pure mathematics, 2.00000 = 2. But not in physics! (This is the sort of thing that drove the purist G.H. Hardy to write that mathematicians are in better contact with reality than are physical scientists.) Those five zeros are important. They indicate how near the exact number 2 we believe the resultant to be. We know the answer is very close to a mathematician's idea of the number 2, but there is an uncertainty of up to ±0.000005. If we chop off the zeros and say simply that $xy = 2$, we allow for an uncertainty of up to ±0.5, and in this case we are entitled to better than that. When we claim a certain number of significant figures, zero gets as much consideration as any other digit.

IN ADDITION AND SUBTRACTION

When measured quantities are added or subtracted, determining the number of significant figures can involve subjective judgment. The best procedure is to expand all the values out to their plain decimal form (if possible), make the calculation as if you were a pure mathematician, and then, at the end of the process, decide how many significant figures you can reasonably claim.

In some cases, the outcome of determining significant figures in a sum or difference is similar to what happens with multiplication or division. Take, for example, the sum $x + y$, where $x = 3.778800 \times 10^{-6}$ and $y = 9.22 \times 10^{-7}$. This calculation proceeds as follows:

$$x = 0.000003778800$$
$$y = 0.000000922$$
$$x + y = 0.0000047008$$
$$= 4.7008 \times 10^{-6}$$
$$= 4.70 \times 10^{-6}$$

But in other instances, one of the values in a sum or difference is insignificant with respect to the other. Let's say that $x = 3.778800 \times 10^{4}$, while $y = 9.22 \times 10^{-7}$. The process of finding the sum goes like this:

$$x = 37,788.00$$
$$y = 0.000000922$$
$$x + y = 37,788.000000922$$
$$= 3.7788000000922 \times 10^{4}$$

In this case, y is so much smaller than x that it doesn't significantly affect the value of the sum. Here, it is best to regard y, in relation to x or to the sum $x +$

y, as the equivalent of a gnat compared with a watermelon. If a gnat lands on a watermelon, the total weight does not appreciably change in practical terms, nor does the presence or absence of the gnat have any effect on the accuracy of the scales. We can conclude that the "sum" here is the same as the larger number. The value y is akin to a nuisance or a negligible error:

$$x + y = 3.778800 \times 10^4$$

A BIT OF CONFUSION

When a value is not in power-of-10 notation, it is best to convert it to that form before deciding on the number of significant figures it contains. If the value begins with 0 followed by a decimal point, for example 0.0004556, it's not too difficult to figure out the number of significant digits (in this case four). But when a number is large, it might not be clear unless the authors tell you what they have in mind.

Have you heard that the speed of light is 300,000,000 meters per second? To how many significant digits do they claim this? One? Two? Three? More? It turns out that this expression is accurate to three significant figures; they mean to say 3.00×10^8 meters per second. A more accurate value is 299,792,000. This expression happens to be accurate to six significant figures: in scientific notation it is 2.99792×10^8.

If this confuses you, you are not alone. It can befuddle the best scientists and engineers. You can do a couple of things to avoid getting into this quagmire of uncertainty. First, always tell your audience how many significant figures you claim when you write an expression as a large number. Second, if you are in doubt about the accuracy in terms of significant figures when someone states or quotes a figure, ask for clarification. It is better to look a little ignorant and get things right, than to act smart and get things wrong.

PROBLEM 7-5
Using a calculator, find the value of sin (0° 0′ 5.33″), rounded off to as many significant figures as are justified.

SOLUTION 7-5
Our angle is specified to three significant figures, so that is the number of significant figures to which we can justify an answer. We are looking for the sine of 5.33 seconds of arc. First, let's convert this angle to degrees. Remember that $1'' = (1/3600)°$. So:

$$\sin (0° \; 0' \; 5.33'') = \sin (5.33/3600)°$$
$$= \sin 0.00148°$$

Using a calculator, we get:

$$\sin 0.00148° = 2.58 \times 10^{-5}$$

PROBLEM 7-6

Suppose a building is 205.55 meters high. The sun is shining down from an angle of 33.5° above the horizon. If the ground near the building is perfectly flat and level, how long is the shadow of the building?

SOLUTION 7-6

The height of the building is specified to five significant figures, but the angle of the sun above the horizon is specified to only three significant figures. Therefore, our answer will have to be rounded to three significant figures.

The situation is illustrated in Fig. 7-2. We assume the building is perfectly flat on top, and that there are no protrusions such as railings or antennas. From the right-triangle model, it is apparent that the height of the building (205.55 meters) divided by the length of the shadow (the unknown, s) is equal to tan 33.5°. Thus:

$$\tan \; 33.5° = 205.55/s$$
$$0.66189 = 205.55/s$$
$$1/s = 0.66189/205.55$$
$$s = 205.55/0.66189 = 310.55$$

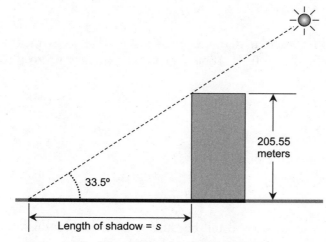

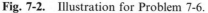

Fig. 7-2. Illustration for Problem 7-6.

This is 311 meters, rounded to three significant figures. When performing this calculation, the tangent of 33.5° was expanded to more than the necessary number of significant figures, and the answer rounded off only at the end. This is in general a good practice, because if rounding is done at early stages in a calculation, the errors sometimes add together and produce a disproportionate error at the end. You might want to try rounding the tangent of 33.5° in the above problem to only three significant figures, and see how, or if, that affects the final, rounded-off answer.

Quiz

Refer to the text in this chapter if necessary. A good score is eight correct. Answers are in the back of the book.

1. What is the sine of 30° 0′ 0″ to three significant figures?
    (a) 0.5
    (b) 0.50
    (c) 0.500
    (d) 0.5000

2. Using a calculator, find the tangent of $(90 - 0.00035675)°$ and round it off to three significant figures in the most unambiguous way possible.
    (a) 161,000
    (b) 160,605
    (c) 1.606×10^5
    (d) 1.61×10^5

3. Suppose the angular diameter of a distant galaxy is measured, and is found to be 0° 30′ 0″ $\pm$ 10%. The error can vary up to plus or minus:
    (a) 5°
    (b) 0.5°
    (c) 0.05°
    (d) 0.005°

4. What is the value of $2 \times 4^2 - 6$?
    (a) 26
    (b) 58
    (c) 20
    (d) There is no way to tell because it is an ambiguous expression

5. Two numbers differ in size by exactly six orders of magnitude. This is a factor of:
   (a) 6
   (b) 6×10
   (c) $10 \times 10 \times 10 \times 10 \times 10 \times 10$
   (d) 2^6

6. Suppose we invent a new unit of angular measure called the flummox (symbol: Fx). We can call 1.00×10^{-9} flummox
   (a) one milliflummox (1.00 mFx)
   (b) one nanoflummox (1.00 nFx)
   (c) one picoflummox (1.00 pFx)
   (d) one kiloflummox (1.00 kFx)

7. What is the value of $\sin^3$ (1), assuming the angle represents precisely one radian, truncated (not rounded off) to three significant figures?
   (a) 0.596
   (b) 0.595
   (c) 8.415×10^{-1}
   (d) 0.841

8. What is the order in which operations should be performed in an expression containing no parentheses?
   (a) Addition, then subtraction, then multiplication, then division, and finally exponentiation
   (b) Exponentiation, then multiplication and division, and finally addition and subtraction
   (c) From left to right
   (d) From the inside out

9. Which of the following expressions might indicate the direction angles of a vector in three-space?
   (a) $(\beta_1, \beta_2, \beta_3)$
   (b) $(\theta^1, \theta^2, \theta^3)$
   (c) $(1_\theta, 2_\theta, 3_\theta)$
   (d) Any of the above

10. What is the product of 8.72×10^5 and 6.554×10^{-5}, taking significant figures into account?
    (a) 57.15088
    (b) 57.151
    (c) 57.15
    (d) 57.2

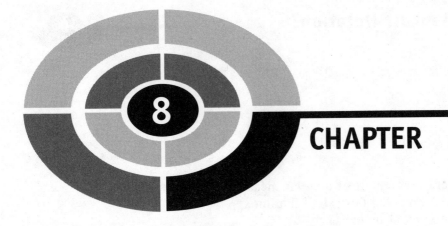

CHAPTER

Surveying, Navigation, and Astronomy

Trigonometry is used to determine distances by measuring angles. In some cases, the angles are exceedingly small, requiring observational apparatus of high precision. In other cases, angle measurement is less critical. Trigonometry can also be used in the reverse sense: determining angles (such as headings or bearings) based on known or measured distances.

Terrestrial Distance Measurement

In order to measure distances using trigonometry, observers rely on a principle of classical physics: rays of light travel in straight lines. This can be taken as "gospel" in surveying and in general astronomy. (There are exceptions to this rule, but they are of concern only to cosmologists and astrophysicists in scenarios where relativistic effects take place.)

PARALLAX

Parallax makes it possible to judge distances to objects and to perceive depth. Figure 8-1 shows the basic principle. Nearby objects appear displaced, relative to a distant background, when viewed with the left eye as compared to the view seen through the right eye. The extent of the displacement depends on the proportional difference between the distance to the nearby object and the distant reference scale, and also on the separation between the left eye and the right eye.

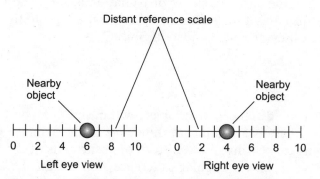

Fig. 8-1.   Parallax allows depth perception; the effect can be used to measure distances.

Parallax can be used for navigation and guidance. If you are heading toward a point, that point seems stationary while other objects seem to move radially outward from it. You can observe this effect while driving down a flat, straight highway. Signs, trees, and other roadside objects appear to move in straight lines outward from a distant point on the road. Parallax simulation gives 3D video games their realism, and is used in stereoscopic imaging.

THE BASE LINE

The use of parallax in distance measurement involves establishing a *base line*. This is a line segment connecting two points of observation. Let's call the observation points P and Q. If the distant object, to which we want to find the distance, is at point R, then we must choose the base line such that $\triangle PQR$ comes as near to being a right triangle as we can manage. We want the base line segment PQ to be perpendicular to either line segment PR or line segment QR, as shown in Fig. 8-2.

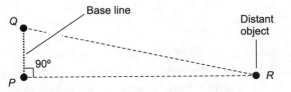

Fig. 8-2. Choosing a base line for distance measurement.

At first thought, getting the base line oriented properly might seem to be a difficult task. But because the distance we want to measure is almost always much longer than the base line, an approximation is good enough. A hiker's compass will suffice to set the base line PQ at a right angle to the line segment connecting the observer and the distant object.

ACCURACY

In order to measure distances to an object within sight, the base line must be long enough so there is a significant difference in the azimuth of the object (that is, its compass bearing) as seen from opposite ends of the base line. By "significant," we mean an angular difference well within the ability of the observing apparatus to detect and measure.

The absolute accuracy (in fixed units such as meters) with which the distance to an object can be measured depends on three factors:

- The distance to the object
- The length of the base line
- The precision of the angle-measuring apparatus

As the distance to the object increases, assuming the base line length stays constant, the absolute accuracy of the distance measurement gets worse; that is, the error increases. As the length of the base line increases, the accuracy improves. As the *angular resolution*, or precision of the angle-measuring equipment, gets better, the absolute accuracy improves, if all other factors are held constant.

PROBLEM 8-1

Suppose we want to determine the distance to an object at the top of a mountain. The base line for the distance measurement is 500.00 meters (which we will call 0.50000 kilometers) long. The angular difference in azimuth is 0.75000° between opposite ends of the base line. How far away is the object?

SOLUTION 8-1

It helps to draw a diagram of the situation, even though it cannot be conveniently drawn to scale. (The base line must be shown out of proportion to its actual relative length.) Figure 8-3 illustrates this scenario. The base line, which is line segment PQ, is oriented at right angles to the line segment PR connecting one end of the base line and the distant object. The right angle is established approximately, using a hiker's compass, but for purposes of calculation, it can be assumed exact, so $\triangle PQR$ can be considered a right triangle.

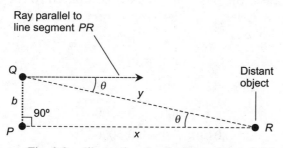

Fig. 8-3.   Illustration for Problems 8-1 and 8-2.

We measure the angle θ between a ray parallel to line segment PR and the observation line segment QR, and find this angle to be 0.75000°. The "parallel ray" can be determined either by sighting to an object that is essentially at an infinite distance, or, lacking that, by using an accurate magnetic compass.

One of the fundamental principles of plane geometry states that pairs of *alternate interior angles* formed by a *transversal* to parallel lines always have equal measure. In this example, we have line PR and an observation ray parallel to it, while line QR is a transversal to these parallel lines. Because of this, the two angles labeled θ in Fig. 8-3 have equal measure.

We use the triangle model for circular functions to calculate the distance to the object. Let b be the length of the base line (line segment PQ), and let x be the distance to the object (the length of line segment PR). Then the following formula holds:

$$\tan \theta = b/x$$

Plugging in known values produces this equation:

$$\tan 0.75000° = 0.50000/x$$
$$0.013090717 = 0.50000/x$$
$$x = 38.195$$

The object on top of the mountain is 38.195 kilometers away from point P.

PROBLEM 8-2
Why can't we use the length of line segment QR as the distance to the object, rather than the length of line segment PR in the above example?

SOLUTION 8-2
We can! Observation point Q is just as valid, for determining the distance, as is point P. In this case, the base line is short compared to the distance being measured. In Fig. 8-3, let y be the length of line segment QR. Then the following formula holds:

$$\sin \theta = b/y$$

Plugging in known values, we get this:

$$\sin 0.75000° = 0.50000/y$$
$$0.013089596 = 0.50000/y$$
$$y = 38.198$$

The percentage difference between this result and the previous result is small. In some situations, the absolute difference between these two determinations (approximately 3 meters) could be of concern, and a more precise method of distance measurement, such as *laser ranging*, would be needed. An example of such an application is precise monitoring of the distance between two points at intervals over a period of time, in order to determine minute movements of the earth's crust along a geological fault line.

STADIMETRY

Stadimetry can be used to measure the distance to an object when the object's height or width is known. The angular diameter of the object is determined by observation. The distance is calculated using trigonometry. This scheme works in the same way as the base-line method described above, except that the "base line" is at the opposite end of the triangle from the observer.

Figure 8-4 shows an example of stadimetry as it might be used to measure the distance d, in meters, to a distant person. Suppose the person's height h,

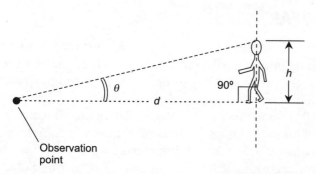

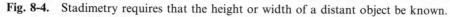

Fig. 8-4. Stadimetry requires that the height or width of a distant object be known.

in meters, is known. The vision system determines the angle θ that the person subtends in the field of view. From this information, the distance d is calculated according to the following formula:

$$d = h/(\tan \theta)$$

In order for stadimetry to be accurate, the linear dimension axis (in this case the axis that depicts the person's height, h) must be perpendicular to a line between the observation point and one end of the object. Also, it is important that d and h be expressed in the same units.

Interstellar Distance Measurement

The distances to stars in our part of the Milky Way galaxy can be measured in a manner similar to the way surveyors measure terrestrial distances. The radius of the earth's orbit around the sun is used as the base line.

THE ASTRONOMICAL UNIT

Astronomers often measure and express interplanetary distances in terms of the *astronomical unit* (AU). The AU is equal to the average distance of the earth from the sun, and is agreed on formally as 1.49597870×10^8 kilometers (this is sometimes rounded off to a figure of 150 million kilometers). The distances to other stars and galaxies can be expressed in astronomical units, but the numbers are large.

THE LIGHT YEAR

Astronomers have invented the *light year*, the distance light travels in one year, to assist in defining interstellar distances so the numbers are reasonable. One light year is the distance a ray of light travels through space in one earth year. You can figure out how far this is by calculation. Light travels approximately 3.00×10^5 kilometers in one second. There are 60 seconds in a minute, 60 minutes in an hour, 24 hours in a day, and about 365.25 days in a year. So a light year is roughly 9.5×10^{12} kilometers.

Let's think on a cosmic scale. The nearest star to our Solar System is a little more than four light years away. The Milky Way, our galaxy, is one hundred thousand (10^5) light years across. The Andromeda galaxy is a little more than two million (2.2×10^6) light years away from our Solar System. Using powerful telescopes, astronomers can peer out to distances of several billion light years (where one billion is defined as 10^9 or one thousand million).

The light year is an interesting unit for expressing the distances to stars and galaxies, but when measurements must be made, it is not the most convenient unit.

THE PARSEC

The true distances to the stars were unknown until the advent of the telescope, with which it became possible to measure extremely small angles. To determine the distances to the stars, astronomers use *triangulation*, the same way surveyors measure distances on the earth.

Figure 8-5 shows how distances to the stars can be measured. This scheme works only for "nearby" stars. Most stars are too far away to produce measurable parallax against a background of much more distant objects, even when they are observed from the earth at different times of the year as it orbits the sun. In Fig. 8-5, the size of the earth's orbit is exaggerated for clarity. The star appears to be in slightly different positions, relative to a background of much more distant objects, at the two observation points shown. The displacement is maximum when the line segment connecting the star and the sun is perpendicular to the line segment connecting the sun with the earth.

Suppose a star thus oriented, and at a certain distance from our Solar System, is displaced by one second of arc when viewed on two occasions, three months apart in time, as shown in Fig. 8-5. When that is the case, the distance between our Solar System and the star is called a *parsec* (a contrac-

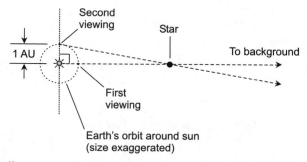

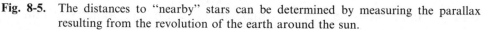

Fig. 8-5. The distances to "nearby" stars can be determined by measuring the parallax resulting from the revolution of the earth around the sun.

tion of "parallax second"). The word "parsec" is abbreviated pc; 1 pc is equivalent to approximately 3.262 light years or 2.063×10^5 AU.

Sometimes units of *kiloparsecs* (kpc) and *megaparsecs* (Mpc) are used to express great distances in the universe. In this scheme, 1 kpc = 1000 pc = 2.063×10^8 AU, and 1 Mpc = 10^6 pc = 2.063×10^{11} AU. Units such as the kiloparsec and the megaparsec make intergalactic distances credible.

The nearest visible object outside our Solar System is the Alpha Centauri star system, which is 1.4 pc away. There are numerous stars within 20 to 30 pc of our sun. The Milky Way is 30 kpc in diameter. The Andromeda galaxy is 670 kpc away. And on it goes, out to the limit of the observable universe, somewhere around 3×10^9 pc, or 3000 Mpc.

PROBLEM 8-3

Suppose we want to determine the distance to a star. We measure the parallax relative to the background of distant galaxies; that background can be considered infinitely far away. We choose the times for our observations so that the earth lies directly between the sun and the star at the time of the first measurement, and a line segment connecting the sun with the star is perpendicular to the line segment connecting the sun with the earth at the time of the second measurement (Fig. 8-6). Suppose the parallax thus determined is 5.0000 seconds of arc ($0° \, 0' \, 5.0000''$). What is the distance to the star in astronomical units?

SOLUTION 8-3

First, consider that the star's distance is essentially the same throughout the earth's revolution around the sun, because the star is many astronomical units away from the sun. We want to find the length of the line segment connecting the sun with the star. This line segment is perpendicular to the line segment connecting the earth with the sun at the time of the second observa-

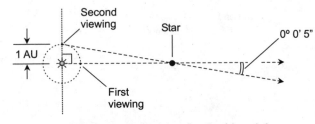

Fig. 8-6.   Illustration for Problem 8-3.

tion. We therefore have a right triangle, and can use trigonometry to find the distance to the star in astronomical units.

The measure of the parallax in Fig. 8-6 is 5.0000 seconds of arc. We divide this by exactly 3600 to get the number of degrees; let's call it $(5/3600)°$ and consider it exact for now. (We'll round the answer off at the end of the calculation.) Let d be the distance from the sun to the star in astronomical units. Then, using the right triangle model:

$$1/d = \tan(5/3600)°$$
$$1/d = 2.4240684 \times 10^{-5}$$
$$d = 41,252.96 \text{ AU}$$

This rounds to 4.1253×10^4 AU because we are justified in going to five significant figures. If you have a good calculator, you can carry out the calculations in sequence without having to write anything down. The display will fill up with a lot of superfluous digits, but you can and should round the answer at the end of the calculation process.

A POINT OF CONFUSION

The parsec can be a confusing unit. If the distance to a star is doubled, then the parallax observed between two observation points, as shown in Fig. 8-5, is cut in half. That does not mean that the number of parsecs to the star is cut in half; it means the number of parsecs is doubled. If taken literally, the expression "parallax second" is a misleading way of expressing the distances to stars, because the smaller the number of parallax seconds, the larger the number of parsecs.

To avoid this confusion, it's best to remember that the parsec is a fixed unit, based on the distance to an object that generates a parallax of one arc second as viewed from two points 1 AU apart. If stadimetry were used in an attempt to measure the distance to a rod 1 AU long and oriented at a right angle to the line of observation (or a person 1 AU tall as shown in Fig. 8-4), then that object would subtend an angle of one arc second as viewed by the observer.

Direction Finding and Radiolocation

Trigonometry can be used to locate an object equipped with a wireless transmitter, based on the azimuth (compass bearing) of that object as observed from two or more widely separated points. Trigonometry can also be used to locate one's own position, based on the signals from wireless transmitters located at two or more widely separated, fixed points.

RADAR

The term *radar* is an acronym derived from the words *radio detection and ranging*. Radio waves having certain frequencies reflect from various objects, especially if those objects contain metals or other electrical conductors. By ascertaining the direction(s) from which radio signals are returned, and by measuring the time it takes for a signal pulse to travel from the transmitter location to a target and back, it is possible to locate flying objects and to evaluate some weather phenomena.

A complete radar set consists of a transmitter, a highly directional antenna, a receiver, and an indicator or display. The transmitter produces microwave pulses that are propagated in a narrow beam. The waves strike objects at various distances. The greater the distance to the target, the longer the delay before the echo is received. The transmitting antenna is rotated so that all azimuth bearings (compass directions) can be observed.

A typical circular radar display is shown in Fig. 8-7. It uses navigator's polar coordinates. The observing station is at the center of the display. Azimuth bearings are indicated in degrees clockwise from true north, and are marked around the perimeter of the screen. The distance, or range, is indicated by the radial displacement of the echo.

FINDING A FOX

A radio receiver, equipped with a signal-strength indicator and connected to a rotatable, directional antenna, can be used to determine the direction from

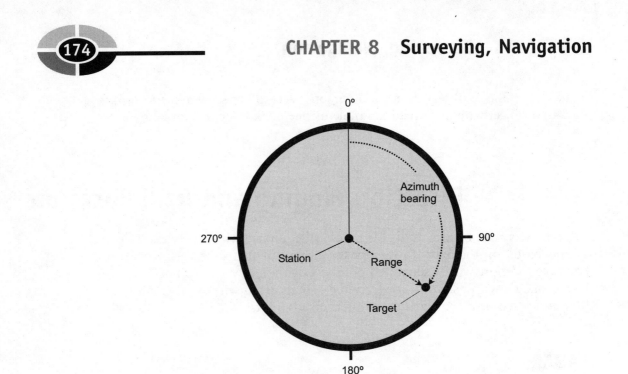

Fig. 8-7.   A radar display shows azimuth and range in navigator's polar coordinates.

which signals are coming. *Radio direction finding* (RDF) equipment aboard a mobile vehicle facilitates determining the location of a signal source. Sometimes hidden transmitters are used to train people in the art of finding signal sources; this is called a "fox hunt."

In an RDF receiver, a loop antenna is generally used. The loop is rotated until a null occurs in the received signal strength. When the null is found, the axis of the loop lies along a line toward the transmitter. When readings are taken from two or more locations separated by a sufficient distance, the transmitter can be pinpointed by finding the intersection point of the azimuth bearing lines on a map or coordinate system. An example is shown in Fig. 8-8. The locations from which readings are taken are indicated by dots labeled X and Y; they form the origins of two navigator's polar coordinate planes. The target, or "fox," is shown by the shaded square. The dashed lines show the azimuth orientations of the tracking antennas at points X and Y. These lines intersect at the location of the "fox."

THE FOX FINDS ITSELF

The captain of a vessel can find the vessel's position by comparing the signals from two fixed stations whose positions are known, as shown in Fig. 8-9. This is, in effect, a "fox hunt in reverse." A vessel, shown by the box, finds its

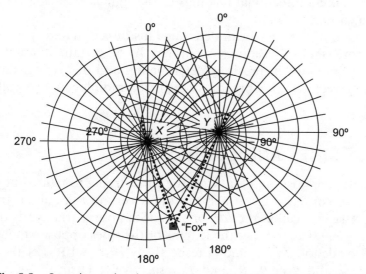

Fig. 8-8. Locating a signal source using radio direction finding (RDF).

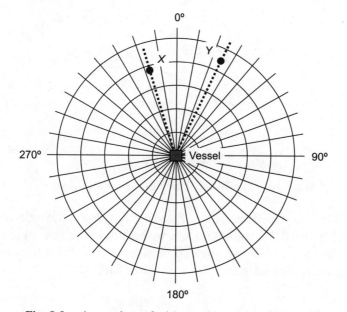

Fig. 8-9. A vessel can find its position using radiolocation.

position by taking directional readings of the signals from sources X and Y. This is called *radiolocation*.

The captain of the vessel can determine his direction and speed by taking two or more sets of readings separated by a certain amount of time. Computers can assist in precisely determining, and displaying, the position and velocity vectors. This process of repeated radiolocation is called *radio-navigation*.

LAWS OF SINES AND COSINES

When finding the position of a target, or when trying to figure out your own location based on bearings, it helps to know certain rules about triangles.

The first rule is called the *law of sines*. Suppose a triangle is defined by three points P, Q, and R. Let the lengths of the sides opposite the vertices P, Q, and R be denoted p, q, and r, respectively (Fig. 8-10). Let the angles at vertices P, Q, and R be θ_p, θ_q, and θ_r, respectively. Then:

$$p/(\sin \theta_p) = q/(\sin \theta_q) = r/(\sin \theta_r)$$

That is to say, the lengths of the sides of any triangle are in a constant ratio relative to the sines of the angles opposite those sides.

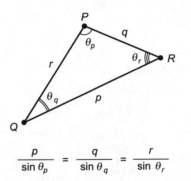

$$\frac{p}{\sin \theta_p} = \frac{q}{\sin \theta_q} = \frac{r}{\sin \theta_r}$$

Fig. 8-10.   The law of sines and the law of cosines.

The second rule is called the *law of cosines*. Suppose a triangle is defined as above and in Fig. 8-10. Suppose you know the lengths of two of the sides, say p and q, and the measure of the angle θ_r between them. Then the length of the third side, r, can be found using the following formula:

$$r = (p^2 + q^2 - 2pq \cos \theta_r)^{1/2}$$

You might recognize this as a modified form of the Pythagorean theorem.

GLOBAL SCENARIOS

The above-mentioned methods of direction finding and radiolocation work well over small geographic regions, within which the surface of the earth (considered as a sphere at sea level) appears nearly flat. But things get more complicated when the radio signals must travel over distances that represent an appreciable fraction of the earth's circumference.

The latitude/longitude system of coordinates, or any other scheme for determining position on the surface of a sphere, involves the use of curved "lines of sight." When trigonometry is used to determine distances and angles on the surface of a sphere, the rules must be modified. We'll look at this in Chapter 11.

PROBLEM 8-4

Suppose a radar set is used to observe an aircraft X and a missile Y, as shown in Fig. 8-11A. Suppose aircraft X is at azimuth 240° 0′ 0″ and range 20.00 kilometers (km), and missile Y is at azimuth 90° 0′ 0″ and range 25.00 km. Suppose both objects are flying directly towards each other, and that aircraft X is moving at 1000 kilometers per hour (km/h) while missile Y is moving at 2000 km/h. How long will it be before the missile and the aircraft collide, assuming neither of them changes course or speed?

SOLUTION 8-4

We must determine the distance, in kilometers, between targets X and Y at the time of the initial observation. This is a made-to-order job for the law of cosines.

Consider the triangle $\triangle XSY$, formed by the aircraft X, the station S, and the missile Y, as shown in Fig. 8-11B. We know that $XS = 20$ km and $SY = 25$ km. We can also deduce that $\angle XSY = 150°$ (the difference between azimuth 240° and azimuth 90°). We now have a triangle with a known angle between two sides of known length. Plugging numbers into the formula for the law of cosines, the distance XY between the aircraft and the missile, in kilometers, is:

$$XY = [20^2 + 25^2 - (2 \times 20 \times 25 \times \cos 150°)]^{1/2}$$
$$= [400 + 625 - (1000 \times -0.8660)]^{1/2}$$
$$= (1025 + 866.0)^{1/2}$$
$$= 1891^{1/2}$$
$$= 43.49$$

Each radial division equals 5 km
Each angular division equals 10°

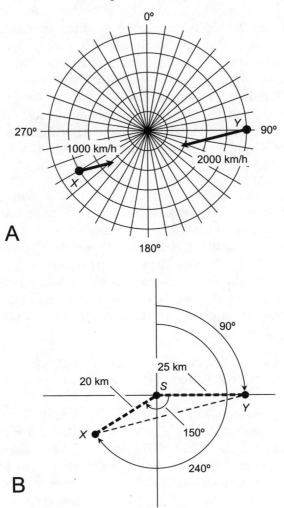

A

B

Fig. 8-11   (A) Illustration for Problem 8-4. (B) Illustration for the solution to Problem 8-4.

The two objects are moving directly towards each other, one at 1000 km/h and the other at 2000 km/h. Their mutual speed is therefore 3000 km/h. If neither object changes course or speed, they will collide after a time t_h, in hours, determined as follows:

$$t_h = (43.49 \text{ km})/(3000 \text{ km/h})$$

$$0.01450 \text{ h}$$

We can obtain the time t_s, in seconds, if we multiply the above by 3600, the number of seconds in an hour:

$$t_s = 0.01450 \times 3600$$
$$= 52.20$$

In a real-life scenario of this sort, a computer would take care of these calculations, and would convey the critical information to the radar operator and the aircraft pilot immediately.

PROBLEM 8-5
Suppose the captain of a vessel wishes to find his location, in terms of latitude and longitude to the nearest minute of arc. He uses direction-finding equipment to measure the azimuth bearings of two buoys, called "Buoy 1" and "Buoy 2," whose latitude and longitude coordinates are known as shown in Fig. 8-12A. The azimuth bearing of Buoy 1 is measured as 350° 0′, and the azimuth bearing of Buoy 2 is measured as 42° 30′, according to the instruments aboard the vessel. What are the latitude and longitude coordinates of the vessel?

SOLUTION 8-5
This problem ought to make you appreciate computers! It's one thing to plot positions on maps, as you've seen in the movies, but it's another thing to manually calculate the values. Computers can do such calculations in a tiny fraction of one second, but it will take us a while longer.

We are working within a geographic region small enough so the surface of the earth can be considered essentially flat, and the lines of longitude can be considered essentially parallel. Therefore, we can convert latitude and longitude to a rectangular coordinate grid with the origin at Buoy 1. Let each minute of arc of latitude or longitude correspond to exactly 1 unit on this grid. Let each axial division on the rectangular coordinate plane equal 10 minutes of arc, as shown in Fig. 8-12B.

We name points P, Q, R, S, T, U, and V, representing intersections among lines and coordinate axes. Lines TU and SV are perpendicular to the horizontal coordinate axis, and line VU is perpendicular to the vertical coordinate axis. We must find either RV or PS, which will let us find the longitude of the vessel relative to Buoy 1, and either PR or SV, which will let us find the latitude of the vessel relative to Buoy 1.

During calculation, let's consider all values exact, and round the answer off when we have found it. Based on the information given, $\angle RPV = 10°$. We

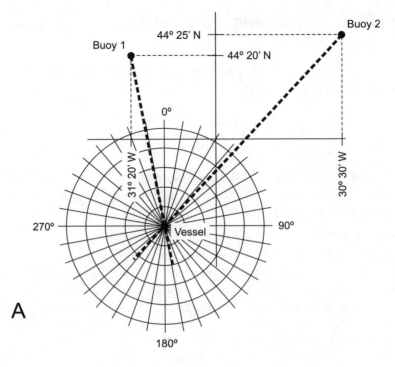

A

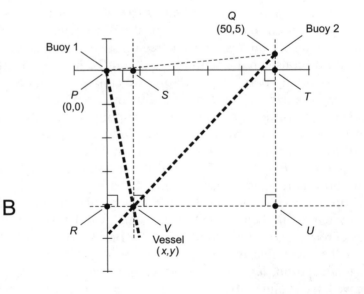

Each horizontal or vertical division equals 10 units

B

Fig. 8-12 (A) Illustration for Problem 8-5. (B) Illustration for the solution to Problem 8-5.

know that $PT = 50$ (units) and $TQ = 5$. We know that $\triangle PTQ$ is a right triangle, so we can use trigonometry to calculate the measure of $\angle TPQ$:

$$\tan \angle TPQ = 5/50 = 0.1$$
$$\angle TPQ = \arctan 0.1 = 5.71059°$$

Because $\angle RPV = 10°$, we can deduce that $\angle VPT = 90° - 10° = 80°$. Because $\angle TPQ = 5.71059°$, we know that $\angle VPQ = 80° + 5.71059° = 85.71059°$. We now know the measure of one of the interior angles of $\triangle VPQ$, an important triangle in the solution of this problem.

Now let's find $\angle PVQ$. This is the angle between the azimuth bearings obtained by the captain of the vessel, or $10° + 42° \, 30'$. Remember that $30' = 0.5°$; therefore $\angle PVQ = 10° + 42.5° = 52.5°$. From this it's easy to figure out $\angle VQP$. It is $180°$ minus the sum of $\angle PVQ$ and $\angle VPQ$:

$$\angle VQP = 180° - (52.5° + 85.71059°)$$
$$= 180° - 138.21059°$$
$$= 41.78941°$$

Next, we can find the distance PQ using the Pythagorean theorem, because $\triangle PTQ$ is a right triangle. We know $PT = 50$ and $TQ = 5$, and also that PQ is the hypotenuse of the triangle. Therefore:

$$PQ = (50^2 + 5^2)^{1/2}$$
$$= (2500 + 25)^{1/2}$$
$$= 2525^{1/2}$$
$$= 50.2494$$

Next, we find the distance PV by applying the law of sines to $\triangle PVQ$:

$$PV/(\sin \angle VQP) = PQ/(\sin \angle PVQ)$$
$$PV = PQ(\sin \angle VQP)/(\sin \angle PVQ)$$
$$= 50.2494(\sin 41.78941°)/(\sin 52.5°)$$
$$= 50.2494 \times 0.6663947/0.7933533$$
$$= 42.2081$$

We now know one of the sides, and all the interior angles, of $\triangle PRV$, which is a right triangle. We can use either $\angle RPV$ or $\angle RVP$ as the basis for finding PR and RV. Suppose we use $\angle RPV$, which measures $10°$. Then:

$$\cos\ 10° = PR/PV$$
$$PR = PV\ \cos 10°$$
$$= 42.2081 \times 0.98481 = 41.5670$$

$$\sin\ 10° = RV/PV$$
$$RV = PV\ \sin 10°$$
$$= 42.2081 \times 0.17365 = 7.3294$$

The final step involves converting these units back into latitude and longitude. Keep in mind that north latitude increases from the bottom of the page to the top, but west longitude increases from the right to the left. Also remember that there are 60 arc minutes in one angular degree.

Let V_{lat} represent the latitude of the vessel. Subtract the displacement PR from the latitude of Buoy 1 and round off to the nearest minute of arc:

$$V_{lat} = 44°\ 20'\ N - 41.5670' = 43°\ 38'\ N$$

Let V_{lon} represent the longitude of the vessel. Subtract the displacement RV (which is the same as PS) from the longitude of Buoy 1 and round off to the nearest minute of arc:

$$V_{lon} = 31°\ 20'\ W - 7.3294' = 31°\ 13'\ W$$

This problem is tough, but it could be worse. Imagine how difficult it would be if we were required to take the earth's curvature into account! The task would still be easy, of course, for a computer to do, and that is why the captains of modern oceangoing vessels leave radiolocation and navigation calculations up to their computers.

QUIZ

Refer to the text in this chapter if necessary. A good score is eight correct. Answers are in the back of the book.

1. On a radar display, a target shows up at azimuth 225°. This is
   (a) northeast of the radar station
   (b) southeast of the radar station
   (c) southwest of the radar station
   (d) northwest of the radar station

2. One parsec
    (a) is equal to one second of arc
    (b) is approximately 3.017×10^{13} km
    (c) is the distance light travels in one second
    (d) is a unit of variable length, depending on parallax

3. The law of sines allows us to find, under certain conditions
    (a) unknown lengths of sides of a triangle
    (b) unknown measures of interior angles of a triangle
    (c) the ratio of the length of a triangle's side to an angle opposite it
    (d) more than one of the above

4. Suppose you are observing several targets on a radar screen. You note
    the azimuth bearings and the ranges of each target at a particular
    moment in time. In order to determine the straight-line distances
    between various pairs of targets, you can use
    (a) triangulation
    (b) stadimetry
    (c) the law of cosines
    (d) parallax

5. Suppose two objects in deep space are the same distance apart as the
    earth is from the sun (1.00 AU). If these objects are 1.00 pc away from
    us, and if a straight line segment connecting them is oriented at a right
    angle with respect to our line of sight (Fig. 8-13), what is the approx-
    imate angle θ, in degrees, that the objects subtend relative to an arbi-
    trarily distant background?
    (a) $1.00°$
    (b) $0.0167°$
    (c) $(2.78 \times 10^{-4})°$
    (d) It is impossible to tell without more information

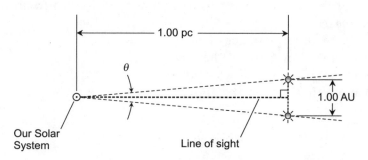

Fig. 8-13.   Illustration for quiz question 5.

6. Suppose you are looking at the echo of an aircraft on a radar display. The radar shows the aircraft is at azimuth 0.000° and range 10.00 km, and is flying on a heading of 90.00°. After a while the aircraft is at azimuth 45.00°. Its range is
   (a) 10.00 km
   (b) 12.60 km
   (c) 14.14 km
   (d) 17.32 km

7. In order to measure distances by triangulation, we must observe the target object from at least
   (a) one reference point
   (b) two reference points
   (c) three reference points
   (d) four reference points

8. If the distance to a star is quadrupled, then the parallax of that star relative to the background of much more distant objects, as observed from two specific, different observation points in the earth's orbit
   (a) becomes half as great
   (b) becomes one-quarter as great
   (c) becomes 1/16 as great
   (d) decreases, but we need more information to know how much

9. As the distance to an object increases and all other factors are held constant, the absolute error (in meters, kilometers, astronomical units, or parsecs) of a distance measurement by triangulation
   (a) increases
   (b) does not change
   (c) decreases
   (d) approaches zero

10. In order to use stadimetry to determine the distance to an object, we must measure
    (a) the angular diameter of the object
    (b) the angular depth of the object
    (c) the number of parsecs to the object
    (d) more than one of the above

Waves and Phase

Trigonometry is important in electricity and electronics, particularly in the analysis of alternating current (AC) and waves.

Alternating Current

In electrical applications, *direct current* (DC) has a *polarity*, or direction, that stays the same over a long period of time. Although the intensity of the current might vary from moment to moment, the electrons always flow in the same direction through the circuit. In *alternating current* (AC), the polarity reverses repeatedly and periodically. The electrons move back and forth; the current ebbs and flows.

PERIOD

In a *periodic AC wave*, the kind discussed in this chapter, the mathematical function of the *amplitude* (the level of current, voltage, power, magnetic-field intensity, or some other variable quantity) versus time repeats precisely and indefinitely. The *period* is the length of time between one repetition of the pattern, or one wave cycle, and the next. This is illustrated in Fig. 9-1 for a simple AC wave.

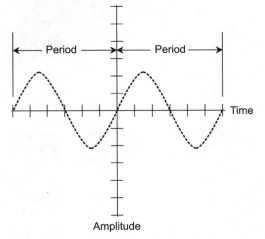

Fig. 9-1. The period of an AC wave is the time it takes for exactly one complete cycle to occur.

The period of an alternating wave can be as short as a minuscule fraction of a second, or as long as thousands of centuries. Some *electromagnetic fields* have periods measured in quadrillionths of a second or smaller. The charged particles held captive by the magnetic field of the sun reverse their direction over periods measured in years, and large galaxies may have magnetic fields that reverse their polarity every few million years. The period of an AC wave, when expressed in seconds, is symbolized T.

FREQUENCY

The *frequency* of an AC wave, denoted f, is the reciprocal of the period. That is, $f = 1/T$, and $T = 1/f$. Prior to the 1970s, frequency was expressed in *cycles per second*, abbreviated cps. High frequencies were expressed in *kilocycles*, *megacycles*, or *gigacycles*, representing thousands, millions, or billions of cycles per second. Nowadays, the standard unit of frequency is known as the *hertz*, abbreviated Hz. Thus, 1 Hz = 1 cps, 10 Hz = 10 cps, and so on.

Higher frequencies are expressed in *kilohertz* (kHz), *megahertz* (MHz), *gigahertz* (GHz), and *terahertz* (THz). The relationships are:

$$1 \text{ kHz} = 1000 \text{ Hz} = 10^3 \text{ Hz}$$
$$1 \text{ MHz} = 1000 \text{ kHz} = 10^6 \text{ Hz}$$
$$1 \text{ GHz} = 1000 \text{ MHz} = 10^9 \text{ Hz}$$
$$1 \text{ THz} = 1000 \text{ GHz} = 10^{12} \text{ Hz}$$

THE SINE WAVE

In its purest form, alternating current has a *sine-wave*, or *sinusoidal*, nature. The waveform in Fig. 9-1 is a sine wave. Any AC wave that concentrates all of its energy at a single frequency has a perfect sine-wave shape. Conversely, any perfect sine-wave electrical signal contains one, and only one, component frequency.

In practice, a wave can be so close to a sine wave that it looks exactly like the sine function on an oscilloscope, when in reality there are traces of signals at other frequencies present. The imperfections in a signal are often too small to see using an oscilloscope, although there are other instruments that can detect and measure them. Utility AC in the United States has an almost perfect sine-wave shape, with a frequency of 60 Hz.

DEGREES OF PHASE

One method of specifying fractions of an AC cycle is to divide it into 360 equal increments called *degrees of phase*, symbolized ° or deg (but it's okay to write out the whole word "degrees"). The value 0° is assigned to the point in the cycle where the signal level is zero and positive-going. The same point on the next cycle is given the value 360°. Halfway through the cycle is 180°; a quarter cycle is 90°; 1/8 cycle is 45°. This is illustrated in Fig. 9-2.

RADIANS OF PHASE

The other method of specifying fractions of an AC cycle is to divide it into 2π, or approximately 6.2832, equal parts. This is the number of radii of a circle that can be laid end-to-end around the circumference. One *radian of phase*, symbolized rad (although you can write out the whole word "radian"), is equal to about 57.29583°. Physicists use the radian more often than the degree when talking about fractional parts of an AC cycle.

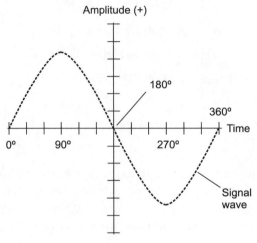

Fig. 9-2.   One cycle of an AC waveform contains 360 degrees of phase.

Sometimes, the frequency of an AC wave is measured in *radians per second* (rad/s), rather than in hertz (cycles per second). Because there are 2π radians in a complete cycle of $360°$, the *angular frequency* of a wave, in radians per second, is equal to 2π times the frequency in hertz. Angular frequency is symbolized by the lowercase, italicized Greek letter omega (ω). Angular frequency can also be expressed in *degrees per second* (deg/s or °/s). The angular frequency of a wave in degrees per second is equal to 360 times the frequency in hertz, or 57.29583 times the angular frequency in radians per second.

INSTANTANEOUS AMPLITUDE

In a sine wave, the amplitude varies with time, over the course of one complete cycle, according to the sine of the number of degrees or radians measured from the start of the wave cycle, or the point on the wave where the amplitude is zero and positive-going.

If the maximum amplitude, also called the *peak amplitude*, that a wave X attains is x_{pk} units (volts, amperes, or whatever), then the *instantaneous amplitude*, denoted x_i, at any instant of time is:

$$x_i = x_{pk} \sin \phi$$

where ϕ is the number of degrees or radians between the start of the cycle and the specified instant in time.

PROBLEM 9-1
What is the angular frequency of household AC in radians per second? Assume the frequency of utility AC is 60.0 Hz.

SOLUTION 9-1
Multiply the frequency in hertz by 2π. If this value is taken as 6.2832, then the angular frequency is:

$$\omega = 6.2832 \times 60.0 = 376.992 \text{ rad/s}$$

This should be rounded off to 377 rad/s, because our input data is given only to three significant figures.

PROBLEM 9-2
A certain wave has an angular frequency of 3.8865×10^5 rad/s. What is the frequency in kilohertz? Express the answer to three significant figures.

SOLUTION 9-2
To solve this, first find the frequency in hertz. This requires that the angular frequency, in radians per second, be divided by 2π, which is approximately 6.2832. The frequency f_{Hz} is therefore:

$$f_{Hz} = (3.8865 \times 10^5)/6.2832$$
$$= 6.1855 \times 10^4 \text{ Hz}$$

To obtain the frequency in kilohertz, divide by 10^3, and then round off to three significant figures:

$$f_{kHz} = .1855 \times 10^4/10^3$$
$$= 61.855 \text{ kHz}$$
$$= 61.9 \text{ kHz}$$

Phase Angle

Phase angle is an expression of the displacement between two waves having identical frequencies. There are various ways of defining this. Phase angles are usually expressed as values ϕ such that $0° \leq \phi < 360°$. In radians, that range is $0 \leq \phi < 2\pi$. Once in a while, you will hear about phase angles specified over a range of $-180° < \phi \leq +180°$. In radians, that range is $-\pi < \phi \leq +\pi$. Phase angle, also called *phase difference*, can be defined only for pairs of waves whose frequencies are the same.

PHASE COINCIDENCE

Two waves are in *phase coincidence* if and only if they have the same frequency and each cycle begins at exactly the same instant in time. Graphically, waves in phase coincidence appear "lined up." This is shown in Fig. 9-3 for two waves having different amplitudes. (If the amplitudes were the same, you would see only one wave.) The phase difference in this case is 0°.

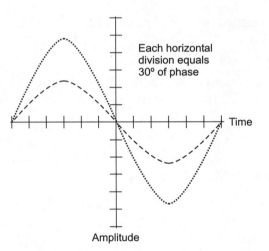

Fig. 9-3.   Two waves in phase coincidence. Graphically, they follow each other along.

If two sine waves are in phase coincidence, the peak amplitude of the *resultant wave*, which is also a sine wave, is equal to the sum of the peak amplitudes of the two *composite waves*. The phase of the resultant is the same as that of the composite waves.

PHASE OPPOSITION

When two sine waves have the same frequency and they begin exactly half a cycle, or 180°, apart, they are said to be in *phase opposition*. This is illustrated in Fig. 9-4. If two sine waves have the same amplitude and are in phase opposition, they cancel each other out because the instantaneous amplitudes of the two waves are equal and opposite at every moment in time.

If two sine waves have different amplitudes and are in phase opposition, the peak value of the resultant wave, which is a sine wave, is equal to the difference between the peak values of the two composite waves. The phase of the resultant is the same as the phase of the stronger of the two composite waves.

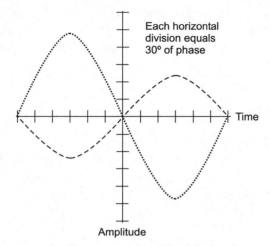

Each horizontal
division equals
30° of phase

Time

Amplitude

Fig. 9-4.   Two waves in phase opposition. Graphically, they are $\frac{1}{2}$ cycle apart.

LEADING PHASE

Suppose there are two sine waves, wave X and wave Y, with identical frequencies. If wave X begins a fraction of a cycle earlier than wave Y, then wave X is said to be *leading* wave Y in phase. For this to be true, X must begin its cycle less than 180° before Y. Figure 9-5 shows wave X leading wave Y by 90°. When one wave leads another, the phase difference can be anything greater than 0° but less than 180°.

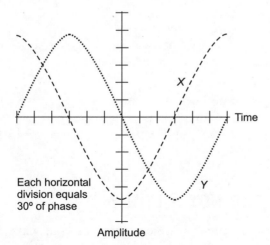

X

Time

Y

Each horizontal
division equals
30° of phase

Amplitude

Fig. 9-5.   Wave X leads wave Y by 90°. Graphically, X appears displaced $\frac{1}{4}$ cycle to the left of (earlier than) Y.

Leading phase is sometimes expressed as a positive phase angle ϕ such that $0° < \phi < +180°$. In radians, this is $0 < \phi < +\pi$. If we say that wave X has a phase of $+\pi/2$ rad relative to wave Y, we mean that wave X leads wave Y by $\pi/2$ rad.

LAGGING PHASE

Suppose wave X begins its cycle more than 180°, but less than 360°, ahead of wave Y. In this situation, it is easier to imagine that wave X starts its cycle later than wave Y, by some value between, but not including, 0° and 180°. Then wave X is *lagging* wave Y. Figure 9-6 shows wave X lagging wave Y by 90°.

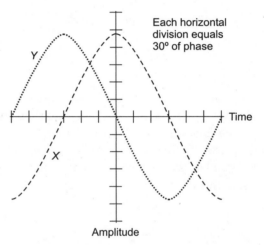

Fig. 9-6.   Wave X lags wave Y by 90°. Graphically, X appears displaced $^{1}/_{4}$ cycle to the right of (later than) Y.

Lagging phase is sometimes expressed as a negative angle ϕ such that $-180° < \phi < 0°$. In radians, this is stated as $-\pi < \phi < 0$. If we say that wave X has a phase of –90° relative to wave Y, we mean that wave X lags wave Y by 90°.

WHEN IS A LEAD NOT A LEAD?

If, while working out a phase problem, you find that wave X differs in phase from wave Y by some angle ϕ that does not fall into the range $-180° < \phi \leq +180°$ ($-\pi < \phi \leq +\pi$ rad), you should reduce the phase difference, either positive or negative, to something that falls in this range. This can be done by adding or subtracting multiples of 360° (2π rad), or by adding or subtracting whole cycles until an acceptable phase difference figure is found.

Suppose, for example, you are told that wave X leads wave Y by exactly 2.75 cycles of phase. That's 2.75 × 360°, or 990°. If you subtract three complete cycles from this, or 3 × 360° = 1080°, you end up with the fact that wave X leads wave Y by –90°. This is the same as saying that wave X lags wave Y by 90°.

VECTOR REPRESENTATIONS OF PHASE

If a sine wave X leads a sine wave Y by ϕ degrees, then the two waves can be drawn as vectors, with vector **X** oriented ϕ degrees counterclockwise from vector **Y**. The waves, when expressed as vectors, are denoted in non-italicized boldface. If wave X lags Y by ϕ degrees, then **X** is oriented ϕ degrees clockwise from **Y**. If two waves are in phase, their vectors overlap (line up). If they are in phase opposition, they point in exactly opposite directions.

The drawings of Fig. 9-7 show four phase relationships between waves X and Y. Wave X always has twice the amplitude of wave Y, so that vector **X** is always twice as long as vector **Y**. At A, wave X is in phase with wave Y. At B, wave X leads wave Y by 90° (π/2 rad). At C, waves X and Y are in phase opposition. In drawing D, wave X lags wave Y by 90° (π/2 rad).

In all cases, with the passage of time, the vectors rotate counterclockwise at the rate of one complete circle per wave cycle. Mathematically, a sine wave is a vector that goes around and around, just like the ball goes around and around your head when you put it on a string and whirl it. The sine wave is a representation of circular motion because the sine function is a circular function.

PROBLEM 9-3
Suppose there are three waves, called X, Y, and Z. Imagine that wave X leads wave Y by 0.5000 rad, while wave Y leads wave Z by precisely 1/8 cycle. By how many degrees does wave X lead or lag wave Z?

SOLUTION 9-3
To solve this, convert all phase angle measures to degrees. One radian is approximately equal to 57.296°. Therefore, 0.5000 rad = 57.296° × 0.5000 = 28.65° (to four significant figures). One-eighth of a cycle is equal to 45.00° (that is 360°/8.000). The phase angles therefore add up, so wave X leads wave Y by 28.65° + 45.00°, or 73.65°.

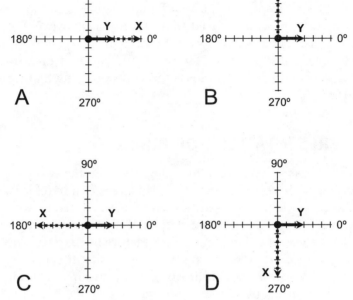

Fig. 9-7. Vector representations of phase difference. At A, wave X is in phase with wave Y. At B, X leads Y by 90°. At C, X and Y are in phase opposition. At D, X lags Y by 90°.

PROBLEM 9-4

Suppose there are three waves X, Y, and Z. Imagine that wave X leads wave Y by 0.5000 rad; wave Y lags wave Z by precisely 1/8 cycle. By how many degrees does wave X lead or lag wave Z?

SOLUTION 9-4

The difference in phase between X and Y in this scenario is the same as that in the previous problem, namely 28.65°. The difference between Y and Z is also the same, but in the opposite sense. Wave Y lags wave Z by 45.00°. This is the same as saying that wave Y leads wave Z by −45.00°. Thus, wave X leads wave Z by 28.65° + (−45.00°), which is equivalent to 28.65° − 45.00° or −16.35°. It is better in this case to say that wave X lags wave Z by 16.35°, or that wave Z leads wave X by 16.35°.

Inductive Reactance

Electrical *resistance*—the extent of the opposition that a medium offers to DC—is a scalar quantity, because it can be expressed on a one-dimensional scale. Resistance is measured in units called *ohms*. Given a certain DC voltage, the electrical current through a device goes down as its resistance goes up. The same law holds for AC through a resistance. A component with resistance has "electrical friction." But in a coil of wire, the situation is more complicated. A coil stores energy as a magnetic field. This makes a coil behave sluggishly when AC is driven through it, as if it has "electrical inertia."

COILS AND CURRENT

If you wind a length of wire into a coil and connect it to a source of DC, the coil becomes warm as energy is dissipated in the resistance of the wire. If the voltage is increased, the current increases also, and the wire gets hot.

Suppose you change the voltage source, connected across the coil, from DC to AC. You vary the frequency from a few hertz (Hz) to many megahertz (MHz). The coil has a certain *inductive reactance* (denoted X_L), so it takes some time for current to establish itself in the coil. As the AC frequency increases, a point is reached at which the current cannot get well established in the coil before the polarity of the voltage reverses. As the frequency is raised, this effect becomes more pronounced. Eventually, if you keep increasing the frequency, the current will hardly get established at all before the polarity of the voltage reverses. Under such conditions, very little current will flow through the coil. Inductive reactance, like resistance, is expressed in ohms. But the "inductive ohm" is a different sort of ohm.

The inductive reactance of a coil (or *inductor*) can vary from zero (a short circuit) to a few ohms (for small coils) to kilohms or megohms (for large coils). Like pure resistance, inductive reactance affects the current in an AC circuit. But unlike pure resistance, inductive reactance changes with frequency. This affects the way the current flows with respect to the voltage.

X_L VS FREQUENCY

If the frequency of an AC source is given (in hertz) as f, and the inductance of a coil is specified (in units called *henrys*) as L, then the inductive reactance (in ohms), X_L, is given by:

$$X_L = 2\pi f L$$

Inductive reactance increases linearly with increasing AC frequency. Inductive reactance also increases linearly with increasing inductance. The value of X_L is directly proportional to f; X_L is also directly proportional to L. These relationships are graphed, in relative form, in Fig. 9-8.

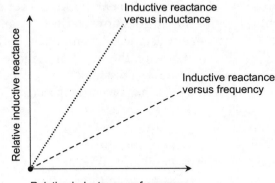

Fig. 9-8.   Inductive reactance increases in a linear manner as the AC frequency goes up, and also as the inductance goes up.

Inductance stores electrical energy as a magnetic field. When a voltage appears across a coil, it takes a while for the current to build up to full value. Thus, when AC is placed across a coil, the current lags the voltage in phase. The current can't keep up with the changing voltage because of the "electrical inertia" in the inductor. Inductive reactance and ordinary resistance combine in interesting ways. Trigonometry can be used to figure out the extent to which the current lags behind the voltage in an inductance–resistance, or *RL*, electrical circuit.

RL PHASE ANGLE

When the resistance in an electronic circuit is significant compared with the inductive reactance, the alternating current resulting from an alternating voltage lags that voltage by less than 90° (Fig. 9-9). If the resistance R is small compared with the inductive reactance X_L, the current lag is almost 90°; as R gets relatively larger, the lag decreases. When R is many times greater than X_L, the phase angle, ϕ_{RL}, is nearly zero. If the inductive reactance vanishes altogether, leaving a pure resistance, then the current and voltage are in phase with each other.

The value of the phase angle ϕ_{RL}, which represents the extent to which the current lags the voltage, can be found using a calculator that has inverse

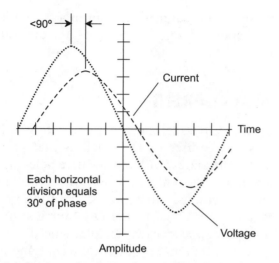

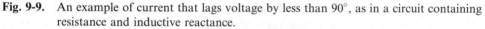

Fig. 9-9.   An example of current that lags voltage by less than 90°, as in a circuit containing resistance and inductive reactance.

trigonometric functions. The angle is the arctangent of the ratio of inductive reactance to resistance:

$$\phi_{RL} = \arctan(X_L/R)$$

PROBLEM 9-5

Find the phase angle between the AC voltage and current in an electrical circuit that has 50 ohms of resistance and 70 ohms of inductive reactance. Express your answer to the nearest whole degree.

SOLUTION 9-5

Use the above formula to find ϕ_{RL}, setting $X_L = 70$ and $R = 50$:

$$\phi_{RL} = \arctan(70/50)$$

$$= \arctan 1.4$$

$$= 54°$$

Capacitive Reactance

Inductive reactance has its counterpart in the form of *capacitive reactance*, denoted X_C. In many ways, inductive and capacitive reactance are alike.

They're both forms of "electrical inertia." But in a capacitive reactance, the voltage has trouble keeping up with the current—the opposite situation from inductive reactance.

CAPACITORS AND CURRENT

Imagine two gigantic, flat, parallel metal plates, both of which are excellent electrical conductors. If a source of DC, such as that provided by a large battery, is connected to the plates (with the negative pole on one plate and the positive pole on the other), current begins to flow immediately as the plates begin to charge up. The voltage difference between the plates starts out at zero and builds up until it is equal to the DC source voltage. This voltage buildup always takes some time, because the plates need time to become fully charged. If the plates are small and far apart, the charging time is short. But if the plates are huge and close together, the charging time can be considerable. The plates form a *capacitor*, which stores energy in the form of an electric field.

Suppose the current source connected to the plates is changed from DC to AC. Imagine that you can adjust the frequency of this AC from a few hertz to many megahertz. At first, the voltage between the plates follows almost exactly along as the AC polarity reverses. As the frequency increases, the charge, or voltage between the plates, does not have time to get well established with each current cycle. When the frequency becomes extremely high, the set of plates behaves like a short circuit.

Capacitive reactance is a quantitative measure of the opposition that a capacitor offers to AC. It, like inductive reactance, varies with frequency and is measured in ohms. But X_C is, by convention, assigned negative values rather than positive values. For any given capacitor, the value of X_C increases negatively as the frequency goes down, and approaches zero from the negative side as the frequency goes up.

X_C VS FREQUENCY

Capacitive reactance behaves, in some ways, like a mirror image of inductive reactance. In another sense, X_C is an extension of X_L into negative values. If the frequency of an AC source is given (in hertz) as f, and the value of a capacitor is given (in units called *farads*) as C, then the capacitive reactance (in ohms), X_C, can be calculated using this formula:

$$X_C = -1/(2\pi f C)$$

Capacitive reactance varies inversely with the negative of the frequency. The function of X_C versus f appears as a curve when graphed, and this curve "blows up negatively" (or, if you prefer, "blows down") as the frequency nears zero. Capacitive reactance varies inversely with the negative of the capacitance, given a fixed frequency. Therefore, the function of X_C versus C also appears as a curve that "blows up negatively" as the capacitance approaches zero. Relative graphs of these functions are shown in Fig. 9-10.

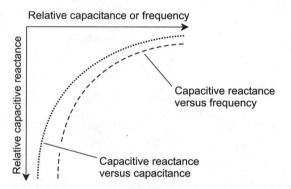

Fig. 9-10.   Capacitive reactance varies inversely with the negative of the frequency. It also varies inversely with the negative of the capacitance, given a fixed frequency.

RC PHASE ANGLE

When the resistance R in an electrical circuit is significant compared with the absolute value (or negative) of the capacitive reactance, the alternating voltage resulting from an alternating current lags that current by less than $90°$. More often, it is said that the current leads the voltage (Fig. 9-11). If R is small compared with the absolute value of X_C, the extent to which the current leads the voltage is almost $90°$; as R gets relatively larger, the phase difference decreases. When R is many times greater than the absolute value of X_C, the phase angle, ϕ_{RC}, is nearly zero. If the capacitive reactance vanishes altogether, leaving just a pure resistance, then the current and voltage are in phase with each other.

The value of the phase angle ϕ_{RC}, which represents the extent to which the current leads the voltage, can be found using a calculator. The angle is the arctangent of the ratio of the absolute value of the capacitive reactance to the resistance:

$$\phi_{RC} = \arctan\left(|X_C|/R\right)$$

CHAPTER 9   Waves and Phase

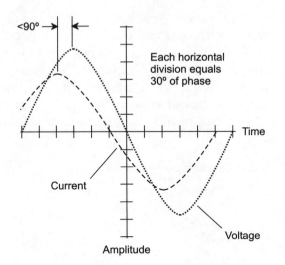

Fig. 9-11.   An example of current that leads voltage by less than 90°, as in a circuit containing
resistance and capacitive reactance.

Because capacitive reactance X_C is always negative or zero, we can also say
this:

$$\phi_{RC} = \arctan(-X_C/R)$$

PROBLEM 9-6
Find the extent to which the current leads the voltage in an AC electronic
circuit that has 96.5 ohms of resistance and –21.1 ohms of capacitive reactance. Express your answer in radians to three significant figures.

SOLUTION 9-6
Use the above formula to find ϕ_{RC}, setting $X_C = -21.1$ and $R = 96.5$:

$$\phi_{RC} = \arctan(|-21.1|/96.5)$$
$$= \arctan(21.1/96.5)$$
$$= \arctan(0.21865)$$
$$= 0.215 \text{ rad}$$

Quiz

Refer to the text in this chapter if necessary. A good score is eight correct. Answers are in the back of the book.

1. Suppose the current lags the voltage in a circuit by 45° of phase. The circuit contains 10 ohms of inductive reactance. How much resistance does the circuit contain? Express the answer to two significant figures.
   (a) 5.0 ohms
   (b) 10 ohms
   (c) 89 ohms
   (d) This question makes no sense because the reactance is capacitive, not inductive

2. Suppose the angular frequency of a wave is specified as 3.14159×10^6 rad/s. What is the period of this wave in seconds? Express the answer to three significant figures.
   (a) 1.59×10^{-7}
   (b) 3.18×10^{-7}
   (c) 2.00×10^{-6}
   (d) 1.00×10^{-6}

3. Approximately how many radians are in a quarter of an AC cycle?
   (a) 0.7854
   (b) 1.571
   (c) 3.142
   (d) 6.284

4. Find the extent to which the current leads the voltage in an AC electronic circuit that has 775 ohms of resistance and 775 ohms of capacitive reactance. Express your answer in degrees to three significant figures.
   (a) 88.7°
   (b) 57.3°
   (c) 45.0°
   (d) None of the above

5. In the stronger wave illustrated by Fig. 9-12, what fraction of a cycle, in degrees, is represented by one horizontal division?
   (a) 60°
   (b) 90°
   (c) 120°
   (d) This question cannot be answered as stated

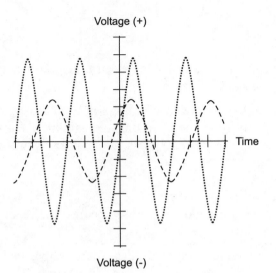

Fig. 9-12.   Illustration for quiz questions 5 and 6.

6. What is the difference in phase between the stronger and weaker waves illustrated in Fig. 9-12?
   (a) 30°
   (b) 45°
   (c) 60°
   (d) This question cannot be answered as stated

7. The peak voltage that appears at a common household utility outlet in the United States is approximately +160 volts. What is the instantaneous voltage three-quarters of the way into a cycle, based on this figure for the peak voltage?
   (a) +40.0 volts
   (b) −61.2 volts
   (c) +113 volts
   (d) −160 volts

8. Suppose there are two sine waves X and Y having identical frequency. Suppose that in a vector diagram, the vector for wave X is 10° counterclockwise from the vector representing wave Y. This means that
   (a) wave X leads wave Y by 10°
   (b) wave X leads wave Y by 170°
   (c) wave X lags wave Y by 10°
   (d) wave X lags wave Y by 170°

9. Suppose there are two sine waves having identical frequency, and their vector representations are at right angles to each other. What is the difference in phase?
   (a) More information is needed to answer this question
   (b) 90°
   (c) 180°
   (d) 2π rad

10. If the angular frequency of a wave is 1000 Hz, then the period of the wave is
    (a) 0.001000 second
    (b) 0.00628 second
    (c) 0.360 second
    (d) impossible to determine because the angular frequency, as stated, makes no sense

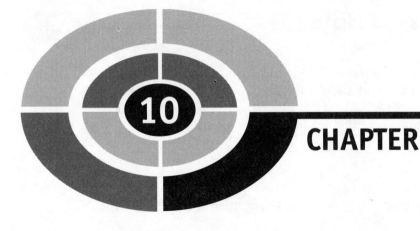

10 CHAPTER

Reflection and Refraction

Trigonometry is used in *optics*, the study of the behavior of light. The phenomena of most interest are *reflection* and *refraction*. A light ray changes direction when it is reflected from a mirror or smooth, shiny surface. If a ray of light passes from one transparent medium into another, the ray may be bent; this is refraction.

Reflection

Any smooth, shiny surface reflects some of the light that strikes it. If the surface is perfectly flat, perfectly shiny, and reflects all of the light that strikes it, then any ray that encounters the surface is reflected away at the same angle at which it hits. You have heard the expression, "The *angle of incidence* equals the *angle of reflection*." This principle, known as the *law of reflection*, is illustrated in Fig. 10-1.

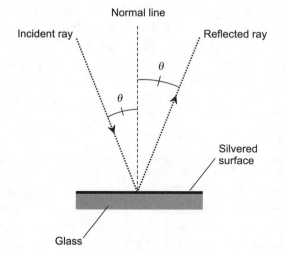

Fig. 10-1. The angle of incidence equals the angle of reflection.

FLAT SURFACES

In optics, the angle of incidence and the angle of reflection are conventionally measured relative to a line normal (perpendicular) to the surface at the point where reflection takes place. In Fig. 10-1, these angles are denoted θ, and can range from $0°$, where the light ray strikes at a right angle with respect to the surface, to almost $90°$, a grazing angle relative to the surface. Sometimes the angle of incidence and the angle of reflection are expressed relative to the surface itself, rather than relative to a normal line.

NON-FLAT SURFACES

If the reflective surface is not perfectly flat, then the law of reflection still applies for each ray of light striking the surface at a specific point. In such a case, the reflection is considered relative to a line normal to a flat plane passing through the point, tangent to the surface at that point. When many parallel rays of light strike a curved or irregular reflective surface at many different points, each ray obeys the law of reflection, but the reflected rays do not all emerge parallel. In some cases they converge; in other cases they diverge. In still other cases the rays are haphazardly scattered.

PROBLEM 10-1
Imagine a room that measures exactly 5.000 meters square, with one mirrored wall. Suppose you stand near one wall (call it "wall W" as shown in

Fig. 10-2A), and hold a flashlight so its bulb is 1.000 meter away from wall W and 3.000 meters away from the mirrored wall. Suppose you aim the flashlight horizontally at the mirrored wall so the center of its beam strikes the mirror at an angle of 70.00° relative to the mirror surface. The beam reflects off the mirror and hits the wall opposite the mirror. The center of the beam strikes the wall opposite the mirror at a certain distance d from wall W. Find d to the nearest centimeter.

SOLUTION 10-1
The path of the light beam is in a plane parallel to the floor and the ceiling, because the flashlight is aimed horizontally. Therefore, we can diagram the situation as shown in Fig. 10-2B. The center of the beam strikes the mirror at an angle of 20.00° relative to the normal. According to the law of reflection, it also reflects from the mirror at an angle of 20.00° relative to the normal. The path of the light beam thus forms the hypotenuses of two right triangles, one whose base measures e meters and whose height is 3.000 meters, and the other whose base measures f meters and whose height is 5.000 meters. If we can determine the values of e and f, then we can easily get the distance d in meters.

Using the right-triangle model for the tangent function, we can calculate e as follows:

$$\tan 20.00° = e/3.000$$
$$0.36397 = e/3.000$$
$$e = 0.36397 \times 3.000 = 1.09191 \text{ meter}$$

We calculate f in a similar way:

$$\tan 20.00° = f/5.000$$
$$0.36397 = f/5.000$$
$$f = 0.36397 \times 5.000 = 1.81985 \text{ meter}$$

Knowing both e and f, we calculate d, in meters, as follows:

$$d = e + f + 1.000$$
$$= 1.09191 + 1.81985 + 1.000$$
$$= 3.91176 \text{ meter}$$

To get d to the nearest centimeter, multiply by 100 and round off:

$$d = 3.91176 \times 100 = 391 \text{ centimeters}$$

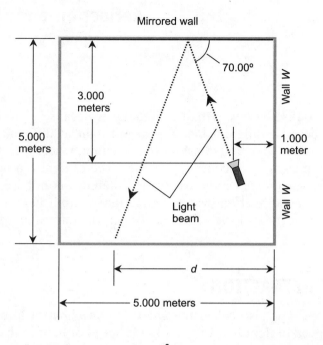

A

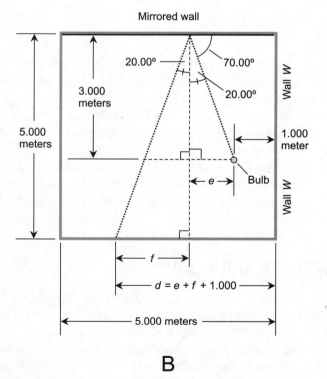

B

Fig. 10-2 (A) Illustration for Problem 10-1. (B) Illustration for the solution to Problem 10-1.

Refraction

A clear pool looks shallower than it actually is because of *refraction*. This effect occurs because different media transmit light at different speeds. The speed of light is absolute and constant in a vacuum, where it travels at about 2.99792×10^8 meters per second. In air, the speed of light is a tiny bit slower than it is in a vacuum; in most cases the difference is not worth worrying about. But in media such as water, glass, quartz, and diamond, the speed of light is significantly slower than it is in a vacuum, and the effects are dramatic.

INDEX OF REFRACTION

The *refractive index*, also called the *index of refraction*, of a medium is the ratio of the speed of light in a vacuum to the speed of light in that medium. If c is the speed of light in a vacuum and c_m is the speed of light in medium M, then the index of refraction for medium M, call it r_m, can be calculated simply:

$$r_m = c/c_m$$

It's important to use the same units, such as meters per second, when expressing c and c_m. According to this definition, the index of refraction of any transparent material is always larger than or equal to 1.

The higher the index of refraction for a transparent substance, the greater the extent to which a ray of light is bent when it strikes the boundary between that substance and air at some angle other than the normal. Various types of glass have different refractive indices. Quartz has a higher refractive index than any glass; diamond has a higher refractive index than quartz. The high refractive index of diamond is responsible for the "sparkle" of diamond "stones."

LIGHT RAYS AT A BOUNDARY

A qualitative example of refraction is shown in Fig. 10-3A, when the refractive index of the first (lower) medium is higher than that of the second (upper) medium. A ray striking the boundary at a right angle (an angle of incidence of 0° relative to the normal) passes through the boundary without changing direction. But a ray that hits at some other angle is bent. The greater the angle of incidence, the sharper the turn the beam takes at the boundary.

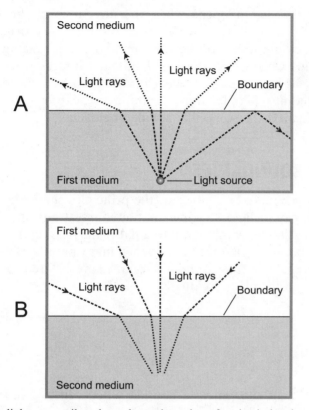

Fig. 10-3. At A, light rays strike a boundary where the refractive index decreases. At B, light rays strike a boundary where the refractive index increases.

When the angle of incidence reaches a certain *critical angle*, then the light ray is not refracted at the boundary, but instead is reflected back into the first medium. This is known as *total internal reflection*.

In air, the speed of light varies just a little bit depending on the density of the gas. Warm air tends to be less dense than cool air, and as a result, warm air has a lower refractive index than cool air. The difference in the refractive index of warm air compared with cooler air can be sufficient to produce total internal reflection if there is a sharp boundary between two air masses whose temperatures are different. This is why, on warm days, you sometimes see "false ponds" over the surfaces of blacktop highways or over stretches of desert sand. This phenomenon is also responsible for certain types of long-distance radio-wave propagation in the earth's atmosphere. Radio waves, like visible light, are electromagnetic in nature, and they obey the rules of reflection and refraction.

Now consider what happens when the directions of the light rays are reversed. This situation is shown in Fig. 10-3B. A ray originating in the first (upper) medium and striking the boundary at a grazing angle is bent downward. This causes distortion of landscape images when viewed from underwater. You have seen this effect if you are a SCUBA diver. The sky, trees, hills, buildings, people, and everything else, can be seen within a circle of light that distorts the scene like a wide-angle lens.

NON-FLAT BOUNDARIES

If the refracting boundary is not flat, the principles shown by Fig. 10-3 still apply for each ray of light striking the boundary at any specific point. The refraction is considered with respect to a flat plane passing through the point, tangent to the boundary at that point. When many parallel rays of light strike a curved or irregular refractive boundary at many different points, each ray obeys the same principle individually.

Snell's Law

When a ray of light encounters a boundary between two substances having different indices (or indexes) of refraction, the extent to which the ray is bent can be determined according to an equation called *Snell's law*.

FROM LOW TO HIGH

Look at Fig. 10-4. Suppose B is a flat boundary between two media M_r and M_s, whose indices of refraction are r and s, respectively. Imagine a ray of light crossing the boundary at point P, as shown. The ray is bent at the boundary whenever the ray does not lie along a normal line, assuming the indices of refraction, r and s, are different.

Suppose $r < s$; that is, the light passes from a medium having a relatively lower refractive index to a medium having a relatively higher refractive index. Let N be a line passing through point P on B, such that N is normal to B at P. Suppose R is a ray of light traveling through M_r that strikes B at P. Let θ be the angle that R subtends relative to N at P. Let S be the ray of light that emerges from P into M_s. Let ϕ be the angle that S subtends relative to N at P. Then line N, ray R, and ray S all lie in the same plane, and $\phi \leq \theta$. (The two

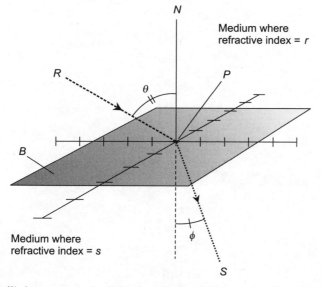

Fig. 10-4.   Snell's law governs the behavior of a ray of light as it strikes a boundary where the index of refraction increases.

angles θ and ϕ are equal if and only if ray R strikes the boundary at an angle of incidence of $0°$, that is, along line N normal to the boundary at point P.) The following equation holds for angles θ and ϕ in this situation:

$$\sin \phi / \sin \theta = r/s$$

The equation can also be expressed like this:

$$s \sin \phi = r \sin \theta$$

FROM HIGH TO LOW

Refer to Fig. 10-5. Again, let B be a flat boundary between two media M_r and M_s, whose absolute indices of refraction are r and s, respectively. In this case imagine that $r > s$; that is, the ray passes from a medium having a relatively higher refractive index to a medium having a relatively lower refractive index. Let N, B, P, R, S, θ, and ϕ be defined as in the previous example. Then line N, ray R, and ray S all lie in the same plane, and $\theta \leq \phi$. (The angles θ and ϕ are equal if and only if R is normal to B.) Snell's law holds in this case, just as in the situation described previously:

$$\sin \phi / \sin \theta = r/s$$

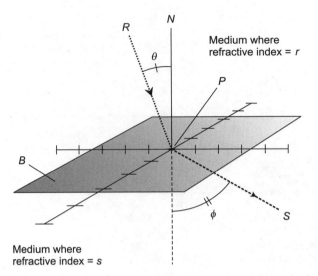

Fig. 10-5. Snell's law for a light ray that strikes a boundary where the index of refraction decreases.

and

$$s \sin \phi = r \sin \theta$$

DETERMINING THE CRITICAL ANGLE

In the situation shown by Fig. 10-5, the light ray passes from a medium having a relatively higher index of refraction, r, into a medium having a relatively lower index, s. Therefore, $s < r$. As angle θ increases, angle ϕ approaches 90°, and ray S gets closer to the boundary plane B. When θ, the angle of incidence, gets large enough (somewhere between 0° and 90°), angle ϕ reaches 90°, and ray S lies exactly in plane B. If angle θ increases even more, ray R undergoes total internal reflection at the boundary plane B. Then the boundary acts like a mirror.

The critical angle is the largest angle of incidence that ray R can subtend, relative to the normal N, without being reflected internally. Let's call this angle θ_c. The measure of the critical angle is the arcsine of the ratio of the indices of refraction:

$$\theta_c = \arcsin (s/r)$$

PROBLEM 10-2

Suppose a laser is placed beneath the surface of a freshwater pond. The refractive index of fresh water is approximately 1.33, while that of air is close to 1.00. Imagine that the surface is perfectly smooth. If the laser is aimed upwards so it strikes the surface at an angle of 30.0° relative to the normal, at what angle, also relative to the normal, will the beam emerge from the surface into the air?

SOLUTION 10-2

Envision the situation in Fig. 10-5 "upside down." Then M_r is the water and M_s is the air. The indices of refraction are $r = 1.33$ and $s = 1.00$. The measure of angle θ is 30.0°. The unknown is the measure of angle ϕ. Use the equation for Snell's law, plug in the numbers, and solve for ϕ. You'll need a calculator. Here's how it goes:

$$\sin \phi / \sin \theta = r/s$$
$$\sin \phi / (\sin 30.0°) = 1.33/1.00$$
$$\sin \phi / 0.500 = 1.33$$
$$\sin \phi = 1.33 \times 0.500 = 0.665$$
$$\phi = \arcsin 0.665 = 41.7°$$

PROBLEM 10-3

What is the critical angle for light rays shining upwards from beneath a freshwater pond?

SOLUTION 10-3

Use the formula for critical angle, and envision the scenario of Problem 10-2, where the laser angle of incidence, θ, can be varied. Plug in the numbers to the equation for critical angle, θ_c:

$$\theta_c = \arcsin (s/r)$$
$$= \arcsin (1.00/1.33)$$
$$= \arcsin 0.752$$
$$= 48.8°$$

Remember that the angles in all these situations are defined with respect to the normal to the surface, not with respect to the plane of the surface.

PROBLEM 10-4

Suppose a laser is placed above the surface of a smooth, freshwater pool that is of uniform depth everywhere, and aimed downwards so the light ray strikes the surface at an angle of 28° relative to the plane of the surface. At what angle, relative to the plane of the pool bottom, will the light beam strike the bottom?

SOLUTION 10-4

This situation is illustrated in Fig. 10-6. The angle of incidence, θ, is equal to $90°$ minus $28°$, the angle at which the laser enters the water relative to the surface. That means $\theta = 62°$. We know that r, the index of refraction of the air, is 1.00, and also that s, the index of refraction of the water, is 1.33. We can therefore solve for the angle ϕ, relative to the normal N to the surface, at which the ray travels under the water:

$$\sin \phi / \sin \theta = r/s$$
$$\sin \phi / \sin 62° = 1.00/1.33$$
$$\sin \phi / 0.883 = 0.752$$
$$\sin \phi = 0.752 \times 0.883 = 0.664$$
$$\phi = \arcsin 0.664 = 42°$$

We're justified to go to two significant figures here, because that is the extent of the accuracy of the angular data we're given. The angle at which the laser travels under the water, relative to the water surface, is $90° - 42°$, or $48°$.

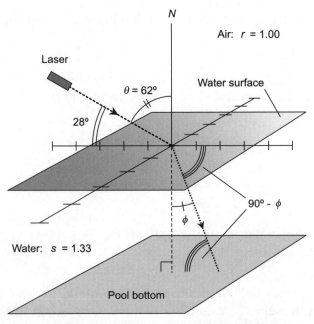

Fig. 10-6. Illustration for Problem 10-4.

Because the pool has a uniform depth, the bottom is parallel to the water surface. Therefore, by invoking the geometric rule for alternate interior angles, we can conclude that the light beam strikes the pool bottom at an angle of 48° with respect to the plane of the bottom.

Dispersion

The index of refraction for a particular substance depends on the wavelength of the light passing through it. Glass, and virtually any other substance having a refractive index greater than 1, slows down light the most at the shortest wavelengths (blue and violet), and the least at the longest wavelengths (red and orange). This variation of the refractive index with wavelength is known as *dispersion*. It is the principle by which a prism works (Fig. 10-7).

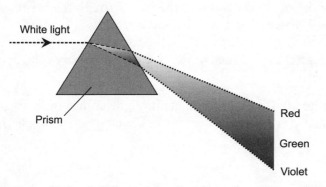

Fig. 10-7.   Light of different colors is refracted at different angles.

RAINBOW SPECTRA

The more a ray of light is slowed down by the glass, the more its path is deflected when it passes through a prism. This is why a prism casts a *rainbow spectrum* when white light passes through it. It is also responsible for the multi-colored glitter of jewelry, especially genuine diamonds, which have high indices of refraction and therefore spread out the colors to a considerable extent.

Dispersion is important in optics for two reasons. First, a prism can be used to make a *spectrometer*, which is a device for examining the intensity of visible light at specific wavelengths. Second, dispersion degrades the quality of white-light images viewed through simple lenses. It is responsible for the "rainbow borders" often seen around objects when viewed through binoculars, telescopes, or microscopes with low-quality lenses.

PROBLEM 10-5

Suppose a ray of white light, shining horizontally, enters a prism whose cross-section is an equilateral triangle and whose base is horizontal (Fig. 10-8A). If the index of refraction of the prism glass is 1.52000 for red light and 1.53000 for blue light, what is the angle δ between rays of red and blue light as they emerge from the prism? Assume the index of refraction of the air is 1.00000 for light of all colors. Find the answer to the nearest hundredth of a degree.

SOLUTION 10-5

There are several ways to approach this problem; all require several steps to complete. Let's do it this way:

- Follow the ray of red light all the way through the prism and determine the angle at which it exits the glass
- Follow the ray of blue light in the same way
- Determine the difference in the two exit angles by subtracting one from the other

Refer to Fig. 10-8B. The ray of white light comes in horizontally, so the angle of incidence is 30° (consider this figure exact). The angle ρ_1 that the ray

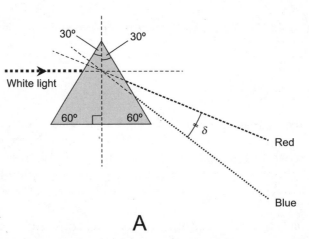

A

Fig. 10-8   (A) Illustration for Problem 10-5.

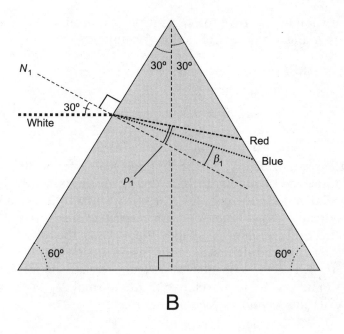

B

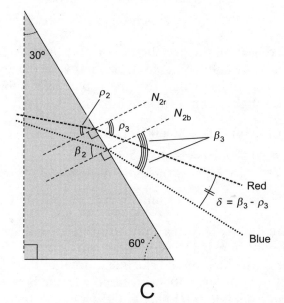

C

Fig. 10-8 (B) Illustration for first part of solution to Problem 10-5. (C) Illustration for second part of solution to Problem 10-5.

of red light subtends relative to N_1, as it passes through the first surface into the glass, is found using the refraction formula:

$$\sin \rho_1 / \sin 30° = 1.00000/1.52000$$
$$\sin \rho_1/0.500000 = 0.657895$$
$$\sin \rho_1 = 0.500000 \times 0.657895 = 0.328948$$
$$\rho_1 = \arcsin 0.328948 = 19.2049°$$

Because the normal line N_1 to the first surface slants at $30°$ relative to the horizontal, the ray of red light inside the prism slants at $30° - 19.2049°$, or $10.7951°$, relative to the horizontal. The line normal to the second surface for the red ray (call it N_{2r}) slants $30°$ with respect to the horizontal, but in the opposite direction from line N_1 (Fig. 10-8C). Thus, the angle of incidence ρ_2, at which the ray of red light strikes the inside second surface of the prism, is equal to $30° + 10.7951°$, or $40.7951°$. Again we use the refraction formula, this time to find the angle ρ_3, relative to the normal N_{2r}, at which the ray of red light exits the second surface of the prism:

$$\sin \rho_3 / \sin 40.7951° = 1.52000/1.00000$$
$$\sin \rho_3/0.653356 = 1.52000$$
$$\sin \rho_3 = 0.653356 \times 1.52000 = 0.993101$$
$$\rho_3 = \arcsin 0.993101 = 83.2659°$$

Now we must repeat all this for the ray of blue light. Refer again to Fig. 10-8B. The ray of white light comes in horizontally, so the angle of incidence is $30°$, as before. The angle β_1 that the ray of blue light subtends relative to N_1, as it passes through the first surface into the glass, is:

$$\sin \beta_1 / \sin 30° = 1.00000/1.53000$$
$$\sin \beta_1/0.500000 = 0.653595$$
$$\sin \beta_1 = 0.500000 \times 0.653595 = 0.326798$$
$$\beta_1 = \arcsin 0.326798 = 19.0745°$$

Because the normal line N_1 to the first surface slants at $30°$ relative to the horizontal, the ray of blue light inside the prism slants at $30° - 19.0745°$, or $10.9255°$, relative to the horizontal. The line normal to the second surface for the blue ray (call it N_{2b}) slants $30°$ with respect to the horizontal, but in the opposite direction from line N_1 (Fig. 10-8C). Thus, the angle of incidence β_2, at which the ray of blue light strikes the inside second surface of the prism, is equal to $30° + 10.9255°$, or $40.9255°$. Again we use the refraction formula, this time to find the angle β_3, relative to the normal N_{2b}, at which the ray of blue light exits the second surface of the prism:

$$\sin \beta_3 / \sin 40.9255° = 1.52000/1.00000$$
$$\sin \beta_3 / 0.655077 = 1.52000$$
$$\sin \beta_3 = 0.655077 \times 1.52000 = 0.995717$$
$$\beta_3 = \arcsin 0.995717 = 84.6952°$$

The difference $\beta_3 - \rho_3$ is the angle δ we seek, the angle between the rays of blue and red light as they emerge from the glass. This, rounded off to the nearest hundredth of a degree, is:

$$\beta_3 - \rho_3 = 84.6952° - 83.2659°$$
$$= 1.43°$$

PROBLEM 10-6

Suppose you want to project a rainbow spectrum onto a screen, so that it measures exactly 10 centimeters (cm) from the red band to the blue band using the prism as configured in Problem 10-5. At what distance d from the screen should the prism be placed? Consider the position of the prism to be the intersection point of extensions of the red and blue rays emerging from the prism, as shown in Fig. 10-9. Consider d to be measured along the blue ray.

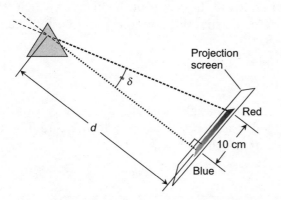

Fig. 10-9.   Illustration for Problem 10-6.

SOLUTION 10-6

This is a simple, straightforward right-triangle problem. The screen is placed so the blue ray is normal to it. We know that $\delta = 1.43°$ (accurate to three significant figures) from the solution to Problem 10-5. We are given that the length of the spectrum from red to blue, as it appears on the screen, is 10 cm, a figure that can be considered exact. We can solve for d as follows:

$$\tan 1.43° = 10/d$$
$$0.024963 = 10/d$$
$$d = 10/0.024963$$
$$d = 400.593$$

This should be rounded off to 401 cm, because we are given the value of δ to only three significant figures.

Quiz

Refer to the text in this chapter if necessary. A good score is eight correct. Answers are in the back of the book.

1. A *step-index optical fiber* has a transparent *core* surrounded by a *cladding*, also transparent, but having a different index of refraction from the core (Fig. 10-10). Suppose that the core of a particular length of optical fiber has a refractive index of 1.45, while the cladding has a refractive index of 1.60, for red light. What is the maximum angle θ, relative to a line along the boundary between the core and the cladding and running parallel to the center of the fiber, at which a ray of red light inside the core can strike the boundary and be reflected back into the core?

  (a) 25°
  (b) 42°
  (c) 65°
  (d) There is no such angle because this fiber is improperly designed. The ray will pass from the core into the cladding no matter what the angle at which it strikes

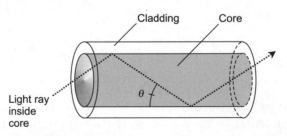

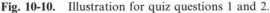
Fig. 10-10.   Illustration for quiz questions 1 and 2.

2. Suppose that the core of the optical fiber shown in Fig. 10-10 has a refractive index of 1.60 while the cladding has a refractive index of 1.45 for red light. What is the maximum angle θ, relative to a line along the boundary between the core and the cladding and running parallel to the center of the fiber, at which a ray of red light inside the core can strike the boundary and be reflected back into the core?
   (a) 25°
   (b) 42°
   (c) 65°
   (d) There is no such angle because this fiber is improperly designed. The ray will pass from the core into the cladding no matter what the angle at which it strikes

3. Suppose a pane of crown glass, with a refractive index of 1.52, is immersed in water, which has a refractive index of 1.33. A ray of light traveling in the water strikes the glass at 45° relative to the normal to the glass surface, and travels through the pane. What angle, relative to the normal, will the ray of light subtend when it leaves the pane and re-enters the water?
   (a) 38°
   (b) 54°
   (c) 45°
   (d) The light will never enter the glass. It will be reflected when it strikes the glass surface

4. Suppose a pane of flint glass, with a refractive index of 1.65, is immersed in water, which has a refractive index of 1.33. A ray of light traveling in the water strikes the glass at 80° relative to the normal, and travels through the pane. What angle, relative to the normal, will the ray of light subtend relative to the normal inside the glass?
   (a) 38°
   (b) 53°
   (c) 65°
   (d) The light will never enter the glass. It will be reflected when it strikes the glass surface

5. Imagine a mineral that is clear, and that has an index of refraction equal to exactly 1.80 for all colors of visible light. A prism made from this mineral will
   (a) produce a well-defined, spread-out rainbow spectrum
   (b) work only underwater
   (c) not bend light rays passing through it
   (d) none of the above

6. Ideal conditions for a "water mirage" over dry land that looks like a pond as a result of reflection exist when the index of refraction of the air near the ground is
   (a) higher than that of the air above it
   (b) lower than that of the air above it
   (c) the same as that of the air above it
   (d) at least as great as the index of refraction of water

7. Imagine the June solstice, when the sun's rays constantly subtend an angle of 23.5° relative to the plane of the earth's equator (Fig. 10-11). Suppose there is a perfectly smooth pond at 43.5° north latitude, and the sun strikes the surface of the pond at high noon. At what angle, relative to a line normal to the surface of the pond, will the sun's rays shine down into the water? Assume the water has a refractive index of 1.33.
   (a) 75°
   (b) 27°
   (c) 63°
   (d) 15°

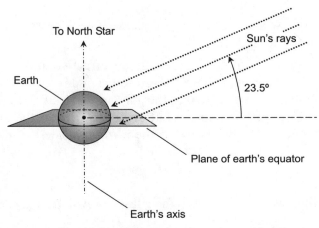

Fig. 10-11. Illustration for quiz question 7.

8. Imagine a solid globe of glass illuminated by monochromatic (single-colored) light, as shown in Fig. 10-12. Suppose that the index of refraction of the glass is 1.55 throughout the globe. Also suppose that the light source is so distant that the rays can be considered parallel, and that the globe is surrounded by air. Inside the globe, the light rays

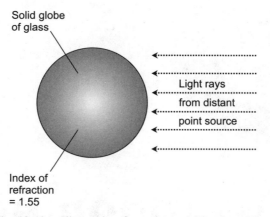

Solid globe
of glass

Light rays
from distant
point source

Index of
refraction
= 1.55

Fig. 10-12. Illustration for quiz questions 8 and 9.

(a) diverge
(b) converge
(c) remain parallel
(d) scatter at random

9. Suppose that the glass globe in Fig. 10-12 is placed in a clear liquid whose index of refraction is 1.75. Suppose the light rays are all parallel inside that liquid medium. Which of the following statements is true?
(a) All of the light rays are reflected from the surface back into the liquid
(b) Some of the rays penetrate into the globe, and some are reflected from the surface back into the liquid
(c) All of the rays penetrate into the globe, and they converge inside it
(d) All of the rays penetrate into the globe, and they diverge inside it

10. What is the critical angle, relative to the normal, of light rays inside a gem whose refractive index is 2.4? Assume the gem is surrounded by air.
(a) 25°
(b) 65°
(c) 67°
(d) 90°

Global Trigonometry

All the trigonometry we've dealt with until now has been on flat surfaces, or in space where all the lines are straight. But in the real world—in particular, on the surface of the earth—lines are not always straight. The route an airline pilot takes to get from New York to Rome is not a straight line; if it were, the aircraft would have to cut through the interior of the planet. The paths of over-the-horizon radio signals are not straight lines. In this chapter, we'll see how trigonometry works on the surface of the earth.

The Global Grid

When the geometric universe is confined to the surface of a sphere, there is no such thing as a straight line or line segment. The closest thing to a straight line in this environment is known as a *great circle* or *geodesic*. The closest thing to a straight line segment is an *arc of a great circle* or *geodesic arc*.

GREAT CIRCLES

The surface of a sphere is the set of all points in space that are equidistant from some center point P. All paths on the surface of a sphere are curved. If

Q and R are any two points on the surface of a sphere, then the straight line segment QR cuts through the interior of the sphere. Navigators and aviators cannot burrow through the earth, and radio waves can't do it either. (Electrical currents at extremely low frequencies can. This is how land-based stations communicate with submarines.)

The shortest distance between any two points Q and R on the surface of the sphere is an arc that lies in a plane passing through P, the center point (Fig. 11-1). The arc QR, representing the shortest on-the-surface distance between the two points, is always part of a great circle, which is a circle on the sphere that has P as its center point. It never fails, as long as the sphere is perfectly round. The surface of the earth, averaged to sea level, is close enough to a perfect sphere that this principle holds quite well. Henceforth in this chapter, when we say "the surface of the earth," it should be understood that we mean "the sphere corresponding to the surface of the earth at sea level." We won't be concerned with local irregularities such as hills, mountains, or buildings.

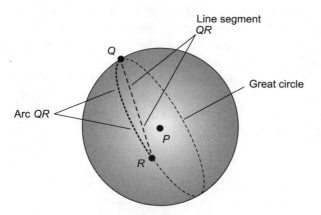

Fig. 11-1.   The shortest distance between points Q and R on a sphere is an arc centered at P, the center of the sphere. Arc QR is always longer than the straight line segment QR.

LATITUDE AND LONGITUDE REVISITED

Latitude and longitude were briefly discussed in Chapter 6. These two terms refer to angles that can be used to uniquely define the position of a point on a sphere, given certain references. Let's review them, and examine them in more detail.

Latitude is defined as an angle, either north (positive) or south (negative), with respect to a great circle representing the equator. The equator is the set of points on the surface of the sphere equidistant from the north geographic pole and the south geographic pole. The geographic poles are the points at which the earth's rotational axis intersects the surface. The latitude, commonly denoted θ, can be as large as $+90°$ or as small as $-90°$, inclusive. That is:

$$-90° \leq \theta \leq +90°$$

or

$$90°S \leq \theta \leq 90°N$$

where "S" stands for "south" and "N" stands for "north."

Longitude is defined as an angle, either east (positive) or west (negative), with respect to a great circle called the *prime meridian*. Longitude is always measured around the equator, or around any circle on the surface of the earth parallel to the equator. The prime meridian has its end points at the north pole and the south pole, and it intersects the equator at a right angle. Several generations ago, it was decided by convention that the town of Greenwich, England, would receive the distinction of having the prime meridian pass through it. For that reason, the prime meridian is also called the *Greenwich meridian*. (When the decision was made, as the story goes, people in France were disappointed, because they wanted the officials to choose the prime meridian so it would pass through Paris. If that had happened, we would be discussing the Paris Meridian right now.) Angles of longitude, denoted ϕ, can range between $-180°$ and $+180°$, not including the negative value:

$$-180° < \phi \leq +180°$$

or

$$180°W < \phi \leq 180°E$$

where "W" stands for "west" and "E" stands for "east."

PARALLELS

For any given angle θ between and including $-90°$ and $+90°$, there is a set of points on the earth's surface such that all the points have latitude equal to θ. This set of points is a circle parallel to the equator; for this reason, all such circles are called *parallels* (Fig. 11-2A). The exceptions are at the extremes $\theta = -90°$ and $\theta = +90°$; these correspond to points at the south geographic pole and the north geographic pole, respectively.

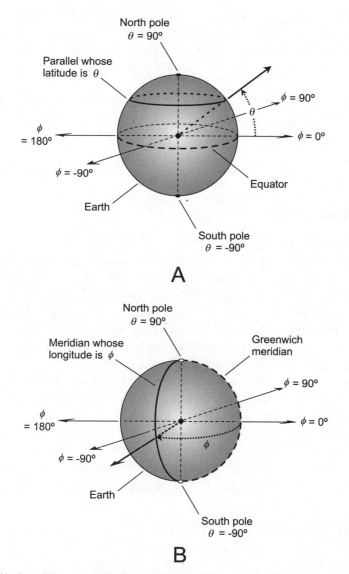

Fig. 11-2 (A) Parallels are circles representing points of equal latitude. Such circles are always parallel to the plane containing the equator. (B) Meridians are half-circles representing points of equal longitude. Such half-circles are always arcs of great circles, with their end points at the north and south geographic poles.

The radius of a given parallel depends on the latitude. When $\theta = 0°$, the parallel is the equator, and its radius is equal to the earth's radius. The earth is not quite a perfect sphere—it is slightly oblate—but it is almost perfect. If we imagine the earth as a perfect sphere with the oblateness averaged out, then we can regard the radius of the earth as equal to 6371 kilometers. That is the radius of the parallel corresponding to $\theta = 0°$. For other values of θ, the radius r (in kilometers) of the parallel can be found according to this formula:

$$r = 6371 \cos \theta$$

The earth's circumference is approximately $6371 \times 2\pi$, or 4.003×10^4 kilometers. Therefore, the circumference k (in kilometers) of the parallel whose latitude is θ can be found using this formula:

$$k = (4.003 \times 10^4) \cos \theta$$

MERIDIANS

For any given angle ϕ such that $-180° < \phi \leq +180°$, there is a set of points on the earth's surface such that all the points have longitude equal to ϕ. This set of points is a half-circle (not including either of the end points) whose center coincides with the center of the earth, and that intersects the equator at a right angle, as shown in Fig. 11-2B. All such open half-circles are called *meridians*. The end points of any meridian, which technically are not part of the meridian, are the south geographic pole and the north geographic pole. (The poles themselves have undefined longitude.)

All meridians have the same radius, which is equal to the radius of the earth, approximately 6371 kilometers. All the meridians converge at the poles. The distance between any particular two meridians, as measured along a particular parallel, depends on the latitude of that parallel. The distance between equal-latitude points on any two meridians ϕ_1 and ϕ_2 is greatest at the equator, decreases as the latitude increases negatively or positively, and approaches zero as the latitude approaches $-90°$ or $+90°$.

DISTANCE PER UNIT LATITUDE

As measured along any meridian (that is, in a north–south direction), the distance $d_{\text{lat-deg}}$ per degree of latitude on the earth's surface is always the same. It can be calculated by dividing the circumference of the earth by 360. If $d_{\text{lat-deg}}$ is expressed in kilometers, then:

$$d_{\text{lat-deg}} = (4.003 \times 10^4)/360 = 111.2$$

The distance $d_{\text{lat-min}}$ per arc minute of latitude (in kilometers) can be obtained by dividing this figure by exactly 60:

$$d_{\text{lat-min}} = 111.2/60.00 = 1.853$$

The distance $d_{\text{lat-sec}}$ per arc second of latitude (in kilometers) is obtained by dividing by exactly 60 once again:

$$d_{\text{lat-sec}} = 1.853/60.00 = 0.03088$$

This might be better stated as $d_{\text{lat-sec}} = 30.88$ meters. That's a little more than the distance between home plate and first base on a major league baseball field.

DISTANCE PER UNIT LONGITUDE

As measured along the equator, the distances $d_{\text{lon-deg}}$ (distance per degree of longitude), $d_{\text{lon-min}}$ (distance per arc minute of longitude), and $d_{\text{lon-sec}}$ (distance per arc second of longitude), in kilometers, can be found according to the same formulas as those for the distance per unit latitude. That is:

$$d_{\text{lon-deg}} = (4.003 \times 10^4)/360 = 111.2$$
$$d_{\text{lon-min}} = 111.2/60.00 = 1.853$$
$$d_{\text{lon-sec}} = 1.853/60.00 = 0.03088$$

These formulas do not work when the east–west distance between any two particular meridians is determined along a parallel other than the equator. In order to determine those distances, the above values must be multiplied by the cosine of the latitude θ at which the measurement is made. Thus, the formulas are modified into the following:

$$d_{\text{lon-deg}} = 111.2\ \cos\ \theta$$
$$d_{\text{lon-min}} = 1.853\ \cos\ \theta$$
$$d_{\text{lon-sec}} = 0.03088\ \cos\ \theta$$

The last formula can be modified for $d_{\text{lon-sec}}$ in meters, as follows:

$$d_{\text{lon-sec}} = 30.88\ \cos\ \theta$$

PROBLEM 11-1

Imagine that a certain large warehouse, with a square floor measuring 100 meters on a side, is built in a community at 60° 0′ 0″ north latitude. Suppose that the warehouse is oriented "kitty-corner" to the points of the compass, so its sides run northeast-by-southwest and northwest-by-southeast. What is the difference in longitude, expressed in seconds of arc, between the west corner and the east corner of the warehouse?

SOLUTION 11-1

The situation is diagrammed in Fig. 11-3. Let $\triangle\phi_{wh}$ be the difference in longitude between the west and east corners of the warehouse. (The $\triangle$ symbol in this context is an uppercase Greek letter delta, which means "the difference in"; it's not the symbol for a geometric triangle.)

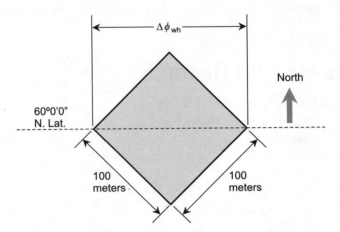

Fig. 11-3.   Illustration for Problems 11-1 and 11-2.

First, we must find the distance in meters between corners of the warehouse. This is equal to $100 \times 2^{1/2}$, or approximately 141.4, meters. Now let's find out how many meters there are per arc second at 60° 0′ 0″ north latitude:

$$d_{\text{lon-sec}} = 30.88 \ \cos \ 60° \ 0' \ 0''$$
$$= 30.88 \times 0.5000$$
$$= 15.44$$

In order to obtain $\triangle\phi_{wh}$, the number of arc seconds of longitude between the east and west corners of the warehouse, we divide 141.4 meters by 15.44 meters per arc second, obtaining:

$$\triangle\phi_{wh} = 141.4/15.44$$
$$= 9.16$$

We round off to three significant figures because that is the extent of the accuracy of our input data (100 meters along each edge of the warehouse). If we want to express this longitude difference in degrees, minutes, and seconds, we write:

$$\triangle\phi_{wh} = 0° \ 0' \ 9.16''$$

PROBLEM 11-2

What is the difference in latitude, expressed in seconds of arc, between the north and the south corners of the warehouse described above?

SOLUTION 11-2

We already know that the distance between corners of the warehouse is 141.4 meters. We also know that there are 30.88 meters of distance per arc second, as measured in a north–south direction, at any latitude. Let $\triangle \theta_{wh}$ be the difference in latitude between the north and the south corners. We divide 141.4 meters by 30.88 meters per arc second, obtaining:

$$\triangle \theta_{wh} = 141.4/30.88$$
$$= 4.58$$

Again, we round off to three significant figures, because that is the extent of the accuracy of our input data (100 meters along each edge of the warehouse). If we want to express this longitude difference in degrees, minutes, and seconds, we write:

$$\triangle \theta_{wh} = 0° \, 0' \, 4.58''$$

Arcs and Triangles

Now we know how latitude and longitude are defined on the surface of the earth, and how to find the differences in latitude and longitude between points along north–south and east–west paths. Let's look at the problem of finding the distance between any two points on the earth, as measured along the arc of a great circle between them.

WHICH ARC?

There are two great-circle arcs that connect any two points on a sphere. One of the arcs goes halfway around the sphere or further, and the other goes halfway around or less. The union of these two arcs forms a complete great circle. The shorter of the two arcs represents the most direct possible route, over the surface of the earth, between the two points.

Let's agree that when we reference the distance between two points on a sphere, we mean to say the distance as measured along the shorter of the two great-circle arcs connecting them. This makes sense. Consider a practical

example. You can get from New York to Los Angeles more easily by flying west across North America than by flying east over the Atlantic, Africa, the Indian Ocean, Australia, and the Pacific Ocean.

SPHERICAL TRIANGLES

A *spherical triangle* is defined by three *vertex* points that all lie on the surface of a sphere. Imagine a triangle on a sphere whose vertex points are Q, R, and S. Let P be the center of the sphere. The spherical triangle, denoted $\triangle_{sph}QRS$, has sides q, r, and s opposite the vertex points Q, R, and S respectively, as shown in Fig. 11-4. (Here, the uppercase Greek delta means "triangle" as in geometry, not "the difference in" as earlier in this chapter!) Each side of the spherical triangle is a great-circle arc spanning less than $360°$. That means that each side must go less than once around the sphere. It is "illegal" to have a spherical triangle with any side that goes all the way around the sphere, or further.

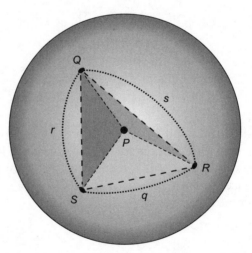

Fig. 11-4.   Each side of a spherical triangle is an arc of a great circle spanning less than $180°$. The vertices of the spherical triangle, along with the center of the sphere itself, define planes as discussed in the text.

For any spherical triangle, there are three ordinary plane triangles defined by the vertices of the spherical triangle and the center of the sphere. In Fig. 11-4, these ordinary triangles are $\triangle PQR$, $\triangle PQS$, and $\triangle PRS$. All three of these triangles define planes in 3D space; call them plane PQR, plane PQS, and plane PRS. Note these three facts:

- Plane *PQR* contains arc *s*
- Plane *PQS* contains arc *r*
- Plane *PRS* contains arc *q*

These concepts and facts are important in defining the interior spherical angles of the spherical triangle $\triangle_{\text{sph}}QRS$.

SPHERICAL POLYGONS

Let's look at the general case, for polygons on the surface of a sphere having three sides or more. A *spherical polygon*, also called a spherical *n-gon*, is defined by *n* vertex points that all lie on the surface of a sphere, where *n* is a whole number larger than or equal to 3. Each side of a spherical *n*-gon is a great-circle arc spanning less than 360°. That means that each side must go less than once around the sphere. It is "illegal" to have a spherical polygon with any side that goes all the way around the sphere, or further.

SPHERICAL ANGLES

The sides of any spherical triangle are curves, not straight lines. The interior angles of a spherical triangle are called *spherical angles*. A spherical angle can be symbolized $\angle_{\text{sph}}$. There are two ways to define this concept.

Definition 1. Consider the three planes defined by the vertices of the spherical triangle and the center of the sphere. In Fig. 11-4, these are plane *PQR*, plane *PQS*, and plane *PRS*. The angles between the arcs are defined like this:

- The angle between planes *PQR* and *PQS*, which intersect in line *PQ*, defines the angle between arcs *r* and *s*, whose vertex is at point *Q*
- The angle between planes *PQR* and *PRS*, which intersect in line *PR*, defines the angle between arcs *q* and *s*, whose vertex is at point *R*
- The angle between planes *PQS* and *PRS*, which intersect in line *PS*, defines the angle between arcs *q* and *r*, whose vertex is at point *S*

How do we express the measure of an angle between two planes, also known as a *dihedral angle*? It's easy to intuit, but hard to explain. Figure 11-5 illustrates the concept. Suppose two planes *X* and *Y* intersect in a common line *L*. Consider line *M* in plane *X* and line *N* in plane *Y*, such that line *M* is perpendicular to line *L*, line *N* is perpendicular to line *L*, and lines *M* and *N* both meet somewhere on line *L*. The angle between the intersecting planes *X* and *Y* can be represented in two ways. The first angle, whose measure is denoted by *u*, is the smaller angle between lines *M* and *N*. The second angle,

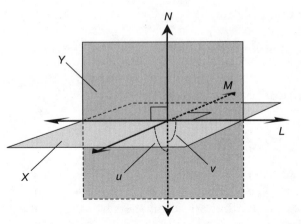

Fig. 11-5. The angle between two intersecting planes X and Y can be represented by u, the acute angle between lines M and N, or by v, the obtuse angle between lines M and N.

whose measure is denoted by v, is the larger angle between lines M and N. The smaller angle is acute, and the larger angle is obtuse. When we talk about the angle at the vertex of a spherical triangle, we must pay attention to whether it is acute or obtuse!

Definition 2. This definition is less rigorous than the first, but it is easier for some people to envision. Let's use a real-world example. On the surface of the earth, suppose two "shortwave" radio signals arrive from two different directions after having traveled partway around the planet along great-circle arcs. If the receiving station uses a directional antenna to check the azimuth bearings (compass directions) of the signals, the curvature of the earth is not considered. The observation is made locally, over a region small enough so that the earth's surface can be considered flat within it. The angle between two great-circle arcs on any sphere that intersect at a point Q can be defined similarly. It is the angle as measured within a circle on the sphere surrounding point Q, such that the circle is so small with respect to the sphere that the circle is essentially a flat disk (Fig. 11-6). Then the great-circle arcs seem to be straight rays running off to infinity, and the angle between them can be expressed as if they lie in the geometric plane tangent to the surface of the sphere at point Q.

ANGULAR SIDES

The sides q, r, and s of the spherical triangle of Fig. 11-4 are often defined in terms of their arc angles ($\angle SPR$, $\angle QPS$, and $\angle RPQ$, respectively), rather than

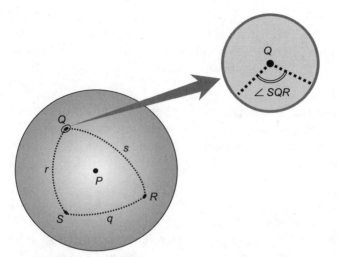

Fig. 11-6. The angle between two intersecting great-circle arcs, as considered within a circular region on the sphere so small that it is essentially a flat disk.

in terms of their actual lengths in linear units. When this is done, it is customary to express the arc angles in radians.

If we know the radius of a sphere (call it r_{sph}), then the length of an arc on the sphere, in the same linear units as we use to measure the radius of the sphere, is equal to the angular measure of the arc (in radians) multiplied by r_{sph}. Suppose we let $|q|$, $|r|$, and $|s|$ represent the lengths of the arcs q, r, and s of $\triangle_{sph}QRS$ in linear units, while their extents in angular radians are denoted q, r, and s. Then the following formulas hold:

$$|q| = r_{sph}q$$
$$|r| = r_{sph}r$$
$$|s| = r_{sph}s$$

In the case of the earth, the linear lengths (in kilometers) of the sides of the spherical triangle $\triangle_{sph}QRS$ are therefore:

$$|q| = 6371q$$
$$|r| = 6371r$$
$$|s| = 6371s$$

PROBLEM 11-3

A great-circle arc on the earth has a measure of 1.500 rad. What is the length of this arc in kilometers?

SOLUTION 11-3

Multiply 6371, the radius of the earth in kilometers, by 1.500, obtaining 9556.5 kilometers. Round this off to 9557 kilometers, because the input data is accurate to four significant figures.

PROBLEM 11-4

Describe and draw an example of a spherical triangle on the surface of the earth in which two interior spherical angles are right angles. Then describe and draw an example of a spherical triangle on the surface of the earth in which all three interior spherical angles are right angles.

SOLUTION 11-4

To solve the first part of the problem, consider the spherical triangle $\triangle_{sph}QRS$ such that points Q and R lie on the equator, and point S lies at the north geographic pole (Fig. 11-7A). The two interior spherical angles $\angle_{sph}SQR$ and $\angle_{sph}SRQ$ are right angles, because sides SQ and SR of $\triangle_{sph}QRS$ lie along meridians, while side QR lies along the equator. (Remember that all of the meridian arcs intersect the equator at right angles.)

To solve the second part of the problem, we construct $\triangle_{sph}QRS$ such that points Q and R lie along the equator and are separated by $90°$ of longitude (Fig. 11-7B). In this scenario, the measure of $\angle_{sph}QSR$, whose vertex is at the north pole, is $90°$. We already know that the measures of angles $\angle_{sph}SQR$ and $\angle_{sph}SRQ$ are $90°$. So all three of the interior spherical angles of $\triangle_{sph}QRS$ are right angles.

THE CASE OF THE EXPANDING TRIANGLE

Imagine what happens to an *equilateral spherical triangle* that starts out tiny and grows larger. (An equilateral spherical triangle has sides of equal angular length and interior spherical angles of equal measure.) An equilateral spherical triangle on the earth that measures 1 arc second on a side is almost exactly the same as a plane equilateral triangle whose sides are 30.88 meters long. The sum of the interior spherical angles, if we measure them with a surveyor's apparatus, appears to be $180°$, and each interior spherical angle appears to be an ordinary angle that measures $60°$. The interior area and the perimeter can be calculated using the formulas used for triangles in a plane.

As the equilateral spherical triangle grows, the measure of each interior spherical angle increases. When each side has a length that is $1/4$ of a great circle (the angular length of each side is $\pi/2$ rad), then each interior spherical angle measures $90°$, and the sum of the measures of the interior spherical angles is three times this, or $270°$. An example is shown in Fig. 11-7B. As the

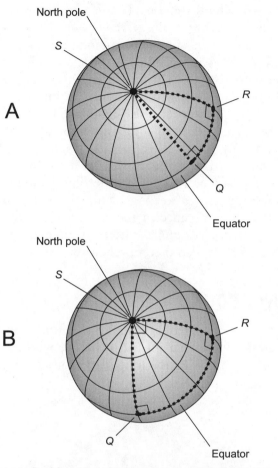

North pole

S

A

R

Q

Equator

North pole

S

B

R

Q

Equator

Fig. 11-7.   Solutions to Problem 11-4. At A, points Q and R are on the equator, and point S is at the north pole. At B, points Q and R are on the equator and are separated by 90° of longitude, while point S is at the north pole.

spherical triangle expands further, it eventually attains a perimeter equal to the circumference of the earth. Then each side has an angular length of $2\pi/3$ rad. The spherical triangle has become a great circle. Its interior area has grown to half the surface area of the planet. The formulas for the perimeter and interior area of a plane triangle do not work for a spherical triangle that is considerable with respect to the size of the globe.

Now think about what happens if the equilateral spherical triangle continues to "grow" beyond the size at which it girdles the earth. The lengths of the sides get shorter, not longer, even though the measures of the interior spherical angles, and the interior area of the spherical triangle, keep increasing. Ultimately, our equilateral spherical triangle becomes so "large" that the three vertices are close together again, perhaps only 1 arc second apart. We have what looks like a triangle similar to the one we started out with—but wait! There are differences. The perimeter is the same, but the interior area is almost that of the whole earth. The inside of this triangle looks like the outside, and the outside looks like the inside. The interior spherical angles are not close to 60°, as they were in the beginning, but instead are close to 300°. They must be measured "the long way around." The sum of their measures is approximately 900°.

This is a bizarre sort of triangle, but in theory, it's entirely "legal." In fact, we can keep on going past a complete circle, letting the interior area and the measures of the interior angles keep growing while the perimeter cycles between zero and the circumference of the earth, over and over. Not many people can envision such a "triangle" after six or eight trips around the world. It's definable, but it's also incomprehensible.

THE LONG WAY AROUND

The foregoing example is not the only instance of "weird spherical triangles" that can be conjured up. Imagine a spherical triangle whose vertices are close to each other, but whose sides go the long way around (Fig. 11-8).

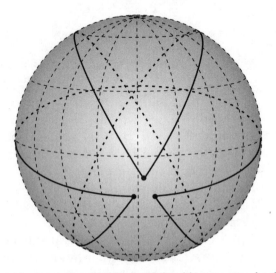

Fig. 11-8.   A spherical triangle in which each of the sides goes nearly all the way around the world.

As long as we are going to be extreme, why stop now? Suppose we free ourselves of the constraint that each side of a spherical n-gon must make less than one circumnavigation of the sphere. Any spherical polygon can then have sides that go more than once around, maybe hundreds of times, maybe millions of times. It's not easy to envision what constitutes the interior of such a monstrosity; we might think of it as a globe wrapped up like a mummy in layer upon layer of infinitely thin tape. And what about the exterior? Perhaps we can think of the mummy-globe again, but this time, wrapped up in infinitely thin tape made of anti-matter.

Mind games like this can be fun, but they reduce to nonsense if taken too seriously. It's a good idea to keep this sort of thing under control, if only for the sake of our sanity. Therefore, when we talk about a spherical polygon, we should insist that its size be limited as follows:

- The perimeter cannot be greater than the circumference of the sphere
- The interior area cannot be greater than half the surface area of the sphere

Any object that violates either of these two criteria should be regarded as "illegal" or "non-standard" unless we are dealing with some sort of exceptional case.

SPHERICAL LAW OF SINES

For any spherical triangle, there is a relationship among the angular lengths (in radians) of the sides and the measures of the interior spherical angles. Let $\triangle_{\mathrm{sph}}QRS$ be a spherical triangle whose vertices are points Q, R, and S. Let the lengths of the sides opposite each vertex point, expressed in radians, be q, r, and s respectively, as shown in Fig. 11-9. Let the interior spherical angles $\angle_{\mathrm{sph}}RQS$, $\angle_{\mathrm{sph}}SRQ$, and $\angle_{\mathrm{sph}}QSR$ be denoted ψ_q, ψ_r, and ψ_s respectively. (The symbol ψ is an italicized, lowercase Greek letter psi; we use this instead of θ to indicate spherical angles.) Then:

$$(\sin q)/(\sin \psi_q) = (\sin r)/(\sin \psi_r) = (\sin s)/(\sin \psi_s)$$

That is to say, the sines of the angular lengths of the sides of any spherical triangle are in a constant ratio relative to the sines of the spherical angles opposite those sides. This rule is known as the *spherical law of sines*. It bears some resemblance to the law of sines for ordinary triangles in a flat plane.

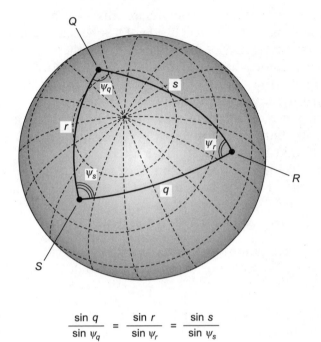

$$\frac{\sin q}{\sin \psi_q} = \frac{\sin r}{\sin \psi_r} = \frac{\sin s}{\sin \psi_s}$$

$$\cos s = \cos q \; \cos r + \sin q \; \sin r \; \cos \psi_s$$

Fig. 11-9.   The law of sines and the law of cosines for spherical triangles.

SPHERICAL LAW OF COSINES

The *spherical law of cosines* is another useful rule for dealing with spherical triangles. Suppose a spherical triangle is defined as above and in Fig. 11-9. Suppose you know the angular lengths of two of the sides, say q and r, and the measure of the spherical angle ψ_s between them. Then the cosine of the angular length of the third side, s, can be found using the following formula:

$$\cos s = \cos q \cos r + \sin q \sin r \cos \psi_s$$

This formula doesn't look much like the law of cosines for ordinary triangles in a flat plane.

EQUILATERAL SPHERICAL TRIANGLE PRINCIPLES

In plane geometry, if a triangle has sides that are all of the same length, then the interior angles are all of the same measure. The converse also holds true:

If the interior angles of a triangle are all of equal measure, then the sides all have the same length.

There is an analogous principle for equilateral triangles on a sphere. If a spherical triangle has sides all of the same angular length, then the interior spherical angles are all of equal measure. The converse is also true: If the interior spherical angles of a spherical triangle all have the same measure, then the angular lengths of the sides are all the same. These principles are important to the solving of the two problems that follow.

PROBLEM 11-5
What are the measures of the interior spherical angles, in degrees, of an equilateral spherical triangle whose sides each have an angular span of 1.00000 rad? Express the answer to the nearest hundredth of a degree.

SOLUTION 11-5
Let's call the spherical triangle $\triangle_{\text{sph}}QRS$, with vertices Q, R, and S, and sides $q = r = s = 1.00000$ rad. Then:

$$\cos q = 0.540302$$
$$\cos r = 0.540302$$
$$\cos s = 0.540302$$
$$\sin q = 0.841471$$
$$\sin r = 0.841471$$

Plug these values into the formula for the law of cosines to solve for $\cos \psi_s$, where ψ_s is the measure of the angle opposite side s. It goes like this:

$$\cos s = \cos q \cos r + \sin q \sin r \cos \psi_s$$
$$0.540302 = 0.540302 \times 0.540302 + 0.841471 \times 0.841471 \times \cos \psi_s$$
$$\cos \psi_s = (0.540302 - 0.291926)/0.708073$$
$$= 0.350777$$

This means that $\psi_s = \arccos 0.350777 = 69.4652°$. Rounding to the nearest hundredth of a degree gives us 69.47°. Because the triangle is equilateral, we know that all three interior spherical angles ψ have the same measure: approximately 69.47°.

PROBLEM 11-6
Suppose we have an equilateral spherical triangle $\triangle_{\text{sph}}QRS$ on the surface of the earth, whose sides each measure 1.00000 rad in angular length, as in the previous problem. Let vertex Q be at the north pole (latitude +90.0000°) and vertex R be at the Greenwich meridian (longitude 0.0000°). Suppose vertex S

is in the western hemisphere, so its longitude is negative. What are the latitude and longitude coordinates of each vertex to the nearest hundredth of a degree?

SOLUTION 11-6
Figure 11-10 shows this situation. We are told that the latitude of point Q (Lat Q) is $+90.0000°$. The longitude of Q (Lon Q) is therefore undefined. We are told that Lon $R = 0.0000°$. Lat R must be equal to $+90.0000°$ minus the angular length of side s. This is $+90.0000° - 1.00000$ rad. Note that 1.00000 rad is approximately equal to $57.2958°$. Therefore:

$$\text{Lat } R = +90.0000° - 57.2958°$$
$$= 32.7042°$$

Rounded off to the nearest hundredth of a degree, Lat $R = +32.70°$. This must also be the latitude of vertex S, because the angular length of side r is the same as the angular length of side s. The longitude of vertex S is equal to the negative of the measure of the interior spherical angle at the pole, or $-\psi$. We know from Solution 11-5 that $\psi = 69.47°$. Therefore, we have these coordinates for the vertices of $\triangle_{\text{sph}}QRS$, rounded off to the nearest hundredth of a degree:

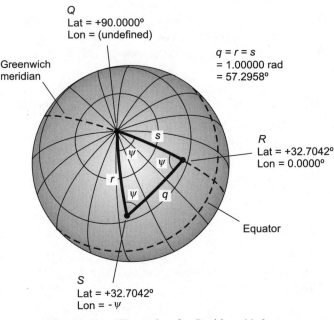

Fig. 11-10. Illustration for Problem 11-6.

$$\text{Lat } Q = +90.00°$$
$$\text{Lon } Q = \text{(undefined)}$$
$$\text{Lat } R = +32.70°$$
$$\text{Lon } R = 0.00°$$
$$\text{Lat } S = +32.70°$$
$$\text{Lon } S = -69.47°$$

Global Navigation

Spherical trigonometry, when done on the surface of the earth, is of practical use for mariners, aviators, aerospace engineers, and military people. It is the basis for determining great-circle distances and headings. Here are four problems relating to global navigation. These problems can be solved almost instantly by computer programs nowadays, but you can get familiar with the principles of global trigonometry by performing the calculations manually.

PROBLEM 11-7
Consider two points R and S on the earth's surface. Suppose the points have the following coordinates:

$$\text{Lat } R = +50.00°$$
$$\text{Lon } R = +42.00°$$
$$\text{Lat } S = -12.00°$$
$$\text{Lon } S = -67.50°$$

What is the angular distance between points R and S, expressed to the nearest hundredth of a radian?

SOLUTION 11-7
To solve this problem, a spherical triangle can be constructed with R and S at two of the vertices, and the third vertex at one of the geographic poles. Let's use the north pole, and call it point Q. (The south pole will work too, but its use is more awkward because the sides of the spherical triangle extend over greater portions of the globe.) We label the sides opposite each vertex q, r, and s. Therefore, q is the angular distance we seek (Fig. 11-11).

We can use the spherical law of cosines to determine q, provided we can figure out three things:

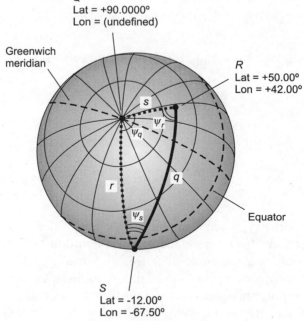

Q
Lat = +90.0000°
Lon = (undefined)

Greenwich
meridian

R
Lat = +50.00°
Lon = +42.00°

s

ψ_r

ψ_q

r

q

ψ_s

Equator

S
Lat = -12.00°
Lon = -67.50°

Fig. 11-11. Illustration for Problems 11-7 through 11-10.

- The measure of ψ_q, the spherical angle at vertex Q
- The angular length of side r
- The angular length of side s

The measure of ψ_q is the absolute value of the difference in the longitudes of the two points R and S:

$$\psi_q = |\text{Lon } R - \text{Lon } S|$$
$$= |+42.00° - (-67.50°)|$$
$$= 42.00° + 67.50°$$
$$= 109.50°$$

The angular length of side r is the absolute value of the difference in the latitudes of points Q and S:

$$r = |\text{Lat } Q - \text{Lat } S|$$
$$= |+90.00° - (-12.00°)|$$
$$= 90.00° + 12.00°$$
$$= 102.00°$$

The angular length of side s is the absolute value of the difference in the latitudes of points Q and R:

$$s = |\text{Lat } Q - \text{Lat } R|$$
$$= |+90.00° - (+50.00°)|$$
$$= 90.00° - 50.00°$$
$$= 40.00°$$

Now that we know r, s, and ψ_q, the spherical law of cosines tells us that:

$$\cos q = \cos r \cos s + \sin r \sin s \cos \psi_q$$

and therefore the following holds:

$$q = \arccos (\cos r \cos s + \sin r \sin s \cos \psi_q)$$
$$= \arccos (\cos 102.00° \times \cos 40.00° + \sin 102.00° \times \sin 40.00° \times \cos 109.50°)$$
$$= \arccos [(-0.20791) \times 0.76604 + 0.97815 \times 0.64279 \times (-0.33381)]$$
$$= \arccos (-0.36915)$$
$$= 1.95 \text{ rad}$$

PROBLEM 11-8
Suppose an aircraft pilot wants to fly a great-circle route from point R to point S in the scenario of Problem 11-7. What is the distance, in kilometers, the aircraft will have to fly? Express the answer to three significant figures.

SOLUTION 11-8
We multiply the angular distance, 1.95 rad, by 6371 kilometers. This gives us 12,400 kilometers, accurate to three significant figures.

PROBLEM 11-9
Suppose an aircraft pilot wants to fly a great-circle route from point R to point S in the scenario of Problem 11-7. What should the initial azimuth heading be as the aircraft approaches cruising altitude after taking off from point R? Express the answer to the nearest degree.

SOLUTION 11-9
In order to determine this, we must figure out the measure of angle ψ_r in degrees. The initial azimuth heading is $360° - \psi_r$. This is because side s of $\triangle_{\text{sph}}QRS$ runs directly northward, or toward azimuth $0°$, from point R. The spherical law of sines tells us this about $\triangle_{\text{sph}}QRS$:

$$(\sin q)/(\sin \psi_q) = (\sin r)/(\sin \psi_r)$$

We can solve this equation for ψ_r by manipulating the above expression and then finding the arcsine:

$$\sin \psi_r = [(\sin r)(\sin \psi_q)]/(\sin q)$$
$$\psi_r = \arcsin \{[(\sin r)(\sin \psi_q)]/(\sin q)\}$$

We already have the following information, having solved Problem 11-7:

$$q = 1.95 \text{ rad}$$
$$\psi_q = 109.5°$$
$$r = 102.00°$$

Plugging in the numbers gives us this:

$$\psi_r = \arcsin \{[(\sin 102.00°)(\sin 109.5°)]/(\sin 1.95 \text{ rad})\}$$
$$= \arcsin [(0.97815 \times 0.94264)/0.92896]$$
$$= \arcsin 0.99255$$
$$= 83.00°$$

This means that the pilot's initial heading, enroute on a great circle from point R to point S, should be $360° - 83.00°$, or $277°$ to the nearest degree. This is $7°$ north of west.

PROBLEM 11-10
Suppose an aircraft pilot wants to fly a great-circle route from point S to point R in the scenario of Problem 11-7. What should the initial azimuth heading be as the aircraft approaches cruising altitude after taking off from point S? Express the answer to the nearest degree.

SOLUTION 11-10
This is similar to the previous problem. We must figure out the measure of angle ψ_s in degrees. The initial azimuth heading is equal to ψ_s because side r of $\triangle_{\text{sph}}QRS$ runs directly northward, or toward azimuth $0°$, from point S. According to the spherical law of sines:

$$(\sin q)/(\sin \psi_q) = (\sin s)/(\sin \psi_s)$$

We can solve this equation for ψ_s by manipulating the above expression and then finding the arcsine:

$$\sin \psi_s = [(\sin s)(\sin \psi_q)]/(\sin q)$$
$$\psi_s = \arcsin \{[(\sin s)(\sin \psi_q)]/(\sin q)\}$$

We already have the following information, having solved Problem 11-7:

$$q = 1.95 \text{ rad}$$
$$\psi_q = 109.5°$$
$$s = 40.00°$$

Plugging in the numbers gives us this:

$$\psi_s = \arcsin \{[(\sin 40.00°)(\sin 109.5°)]/(\sin 1.95 \text{ rad})\}$$
$$= \arcsin [(0.64279 \times 0.94264)/0.92896]$$
$$= \arcsin 0.65226$$
$$= 40.71°$$

This means that the pilot's initial heading, enroute on a great circle from point S to point R, should be $41°$ to the nearest degree. This is $41°$ east of north.

Quiz

Refer to the text in this chapter if necessary. A good score is eight correct. Answers are in the back of the book.

1. When the angular lengths of two sides of a spherical triangle are known, and the measure of the spherical angle between those two sides is also known, then the angular length of the side opposite the known spherical angle can be found using
   (a) the Pythagorean theorem
   (b) latitude and longitude
   (c) addition of angles
   (d) none of the above

2. Suppose the universe is a gigantic sphere with a circumference of 2.4×10^{10} parsecs (pc). How long is a geodesic arc on the surface of that sphere whose angular measure is equal to $1° \, 0' \, 0''$?
   (a) 1.33×10^7 pc
   (b) 6.67×10^7 pc
   (c) 3.82×10^9 pc
   (d) 7.64×10^9 pc

3. Suppose two shortwave radio signals arrive at a receiving station after having traveled thousands of kilometers along geodesic paths. One

signal comes from azimuth 60° and the other comes from azimuth 140°. The difference between these azimuth angles, 80°, is an example of
- (a) the law of sines
- (b) the law of cosines
- (c) global navigation
- (d) a spherical angle

4. Suppose an aircraft pilot wants to fly a great-circle route from point R to point S, where:

$$\text{Lat } R = +45.00°$$
$$\text{Lon } R = -97.00°$$
$$\text{Lat } S = +8.00°$$
$$\text{Lon } S = +55.00°$$

The length of a geodesic (great-circle route) between these two points is approximately
- (a) 10,500 kilometers
- (b) 12,000 kilometers
- (c) 13,500 kilometers
- (d) 17,500 kilometers

5. Consider a four-sided polygon on the surface of the earth, whose vertices Q, R, S, and T are at the following latitudes and longitudes:

$$\text{Lat } Q = +30° \, 0' \, 0''$$
$$\text{Lon } Q = 0° \, 0' \, 0''$$
$$\text{Lat } R = +30° \, 0' \, 0''$$
$$\text{Lon } R = +90° \, 0' \, 0''$$
$$\text{Lat } S = +30° \, 0' \, 0''$$
$$\text{Lon } S = 180° \, 0' \, 0''$$
$$\text{Lat } T = +30° \, 0' \, 0''$$
$$\text{Lon } T = -90° \, 0' \, 0''$$

This is a *spherical square*. The measure of each interior spherical angle of this spherical square is:
- (a) less than 90°
- (b) equal to 90°
- (c) more than 90°
- (d) impossible to determine without more information

6. Suppose a spherical polygon has vertices at the following locations on the earth's surface:

$$\text{Lat } Q = 0° \ 0' \ 0''$$
$$\text{Lon } Q = +24° \ 0' \ 0''$$
$$\text{Lat } R = 0° \ 0' \ 0''$$
$$\text{Lon } R = -110° \ 0' \ 0''$$
$$\text{Lat } S = 0° \ 0' \ 0''$$
$$\text{Lon } S = +134° \ 0' \ 0''$$

The angular length of the great-circle arc QR, to the nearest hundredth of a radian, is
   (a) 4.68 rad
   (b) 2.34 rad
   (c) 0.43 rad
   (d) 0.21 rad

7. The sum of the measures of the interior spherical angles of the spherical polygon described in Question 6, to the nearest degree, is
   (a) 540°
   (b) 312°
   (c) 132°
   (d) 48°

8. Consider a spherical equilateral triangle on the earth's surface with each side measuring exactly 11,000 kilometers in length. What is the measure of each interior spherical angle of this spherical triangle to the nearest degree?
   (a) 101°
   (b) 79°
   (c) 60°
   (d) It cannot be determined without more information

9. Suppose Q and R are two points on the earth that are widely separated. Imagine that a shortwave radio transmitter is located at point Q, and its signal is received at point R after propagating along two great-circle arcs: one arc representing the *short path* and the other arc representing the *long path*. The two signals arrive at point R from
   (a) the same point of the compass
   (b) points of the compass that differ by 90°
   (c) points of the compass that differ by 180°
   (d) all points of the compass simultaneously

10. The sum of the measures of the interior angles of an equilateral spherical triangle is always greater than
    (a) 540°
    (b) 360°
    (c) 180°
    (d) none of the above

Test: Part Two

Do not refer to the text when taking this test. You may draw diagrams or use a calculator if necessary. A good score is at least 38 correct. Answers are in the back of the book. It's best to have a friend check your score the first time, so you won't memorize the answers if you want to take the test again.

1. Fill in the blank: "A kilometer is _____ orders of magnitude larger than a millimeter."
   (a) 0
   (b) 2
   (c) 4
   (d) 6
   (e) 8

2. Suppose an aircraft is detected on radar at azimuth (or bearing) 90°. It flies on a heading directly north, and continues on that heading. As we watch the aircraft on the screen
   (a) its azimuth and range both increase
   (b) its azimuth increases and its range decreases
   (c) its azimuth decreases and its range increases
   (d) its azimuth and range both decrease
   (e) its azimuth and range both remain constant

3. The period of a sine wave contains
   (a) 90° of phase
   (b) 180° of phase
   (c) 270° of phase
   (d) 360° of phase
   (e) none of the above

4. The angular diameter of a distant object (in degrees or radians) can be used to determine the distance to that object, if the actual diameter of the object (in linear units such as kilometers) is known. This technique is called
   (a) parallax
   (b) line sighting
   (c) angle sighting
   (d) surveying
   (e) stadimetry

5. Assume that waves X and Y shown in Fig. Test 2-1 are both sinusoidal. Also assume that the time and amplitude scales are linear. The peak amplitude of wave Y is
   (a) four times the peak amplitude of wave X
   (b) twice the peak amplitude of wave X
   (c) half the peak amplitude of wave X
   (d) a quarter of the peak amplitude of wave X
   (e) the same as the peak amplitude of wave X

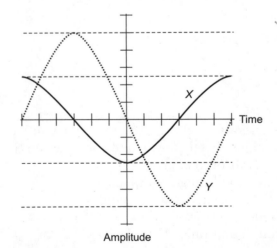

Fig. Test 2-1. Illustration for Questions 5, 6, and 7 in the test for Part Two.

6. Assume that waves X and Y shown in Fig. Test 2-1 are both sinusoidal. Also assume that the time and amplitude scales are linear. The frequency of wave Y is
   (a) four times the frequency of wave X
   (b) twice the frequency of wave X
   (c) half the frequency of wave X
   (d) a quarter of the frequency of wave X
   (e) the same as the frequency of wave X

7. Assume that waves X and Y shown in Fig. Test 2-1 are both sinusoidal. Also assume that the time and amplitude scales are linear. Wave Y
   (a) lags wave X by $\pi/2$ radians of phase
   (b) leads wave X by $\pi/2$ radians of phase
   (c) is in phase coincidence with wave X
   (d) is in phase opposition relative to wave X
   (e) bears no phase relationship to wave X

8. The number 5.33×10^{-4}, written out in full, is
   (a) 5330000
   (b) 53300
   (c) 5.33
   (d) 0.0533
   (e) 0.000533

9. What is the phase difference, in degrees, between the two waves defined by the following functions:

$$y = \sin\ x$$
$$y = -4\ \cos\ x$$

   (a) $0°$
   (b) $45°$
   (c) $90°$
   (d) $180°$
   (e) It is undefined, because the two waves do not have the same frequency

10. What is the phase difference, in degrees, between the two waves defined by the following functions:

$$y = \sin\ x$$
$$y = \cos\ (-4x)$$

(a) 0°
(b) 45°
(c) 90°
(d) 180°
(e) It is undefined, because the two waves do not have the same frequency

11. What is the square root of 29, truncated (not rounded) to two significant figures?
(a) 5.3
(b) 5.38
(c) 5.4
(d) 5.39
(e) None of the above

12. Imagine two alternating-current, sinusoidal waves X and Y that have the same frequency. Suppose wave X leads wave Y by 300° of phase. The more common way of saying this is
(a) wave X leads wave Y by 60°
(b) wave Y leads wave X by 60°
(c) wave Y lags wave X by 300°
(d) the two waves are in phase coincidence
(e) nothing! The described situation is impossible

13. Suppose a surveyor uses triangulation to measure the distances to various objects. As the distance to an object increases, assuming all other factors remain constant, the absolute error, expressed in meters, of the distance measurement
(a) diminishes
(b) stays the same
(c) increases
(d) cannot be determined
(e) is a meaningless expression

14. What is the product of 5.66×10^5 and 1.56999×10^{-3}, rounded to the justifiable number of significant figures?
(a) 8.88
(b) 888
(c) 8.89
(d) 889
(e) None of the above

15. In some transparent materials, the index of refraction depends on the color of the light. This effect is called
    (a) refraction
    (b) total internal reflection
    (c) declination
    (d) dispersion
    (e) distortion

16. The critical angle for light rays that strike a boundary between two transparent substances depends on
    (a) the ratio of the indices of refraction of the substances
    (b) the ratio of the declinations of the substances
    (c) the ratio of the distortions of the substances
    (d) the ratio of the dispersions of the substances
    (e) none of the above

17. Fill in the blank: "The lengths of the sides of any triangle are in a constant ratio relative to the ____ of the angles opposite those sides."
    (a) tangents
    (b) sines
    (c) cosines
    (d) secants
    (e) cotangents

18. Figure Test 2-2 shows the path of a light ray R, which becomes ray S as it crosses a flat boundary B between media having two different indexes of refraction r and s. Suppose that line N is normal to plane B. Also suppose that line N, ray R, and ray S all intersect plane B at point P. If $\theta = 35°$ and $\phi = 60°$, we can conclude that
    (a) $r > s$
    (b) $r = s$
    (c) $r < s$
    (d) the illustrated situation is impossible
    (e) rays R and S cannot lie in the same plane

19. Imagine a light ray R, which becomes ray S as it crosses a flat boundary B between media having two different indexes of refraction r and s, as shown in Fig. Test 2-2. Suppose that line N is normal to plane B. Also suppose that line N, ray R, and ray S all intersect plane B at point P. We are given the following equation relating various parameters in this situation:

$$s \sin \phi = r \sin \theta$$

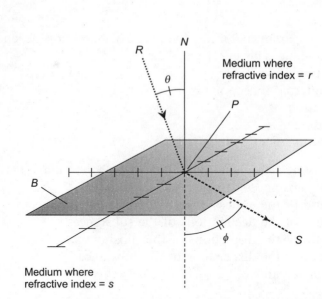

R

N

Medium where
refractive index = r

P

θ

B

φ

S

Medium where
refractive index = s

Fig. Test 2-2.   Illustration for Questions 18, 19, and 20 in the test for Part Two.

Suppose we are told, in addition to all of the above information, that
$\phi = 60°\ 00'$, $\theta = 35°\ 00'$, and $r = 1.880$. From this, we can determine that
  (a) $s = 0.532$
  (b) $s = 1.000$
  (c) $s = 1.245$
  (d) $s = 2.134$
  (e) none of the above

20. Imagine a light ray R, which encounters a flat boundary B between
media having two different indexes of refraction r and s, as shown in
Fig. Test 2-2. Suppose that line N is normal to plane B. Also suppose
that line N and ray R intersect plane B at point P. Suppose we are told
that $r > s$. What can we conclude about the angle of incidence θ at
which ray R undergoes total internal reflection at the boundary plane
B?
  (a) The angle θ must be less than $\pi/2$ rad
  (b) The angle θ must be less than $\pi/3$ rad
  (c) The angle θ must be less than 1 rad
  (d) The angle θ must be less than $\pi/4$ rad
  (e) There is no such angle θ, because no ray R that strikes B as shown
     can undergo total internal reflection if $r > s$

21. In the equation $\theta_1 + \theta_2 = \arctan x$, the variables θ_1 and θ_2 represent
    (a) circular functions
    (b) tangents
    (c) angles
    (d) logarithms
    (e) hyperbolic functions

22. When a light ray passes through a boundary from a medium having an index of refraction r into a medium having an index of refraction s, the critical angle, θ_c, is given by the formula:

$$\theta_c = \arcsin (s/r)$$

What does this formula tell us about rays striking a boundary where $r = 2s$?
    (a) All of the incident rays pass through
    (b) None of the incident rays pass through
    (c) Only the incident rays striking at less than $30°$ relative to the normal pass through
    (d) Only the incident rays striking at more than $30°$ relative to the normal pass through
    (e) The critical angle is not defined if $r = 2s$

23. The shortest distance between two points on the surface of a sphere, as determined over the surface (not cutting through the sphere), is known as
    (a) an arc of a great circle
    (b) a spherical line segment
    (c) a linear sphere segment
    (d) a spherical angle
    (e) a surface route

24. Imagine a point P on a sphere where two different great circles C_1 and C_2 intersect. Now imagine some point Q (other than P) on great circle C_1, and some point R (other than P) on great circle C_2. The angle QPR, as expressed on the surface of the sphere, is an example of
    (a) an acute angle
    (b) a triangular angle
    (c) an obtuse angle
    (d) a circumferential angle
    (e) a spherical angle

25. As you drive along a highway, the compass bearings of nearby objects change more rapidly than the compass bearings of distant objects. This is because of
    (a) stadimetry
    (b) parsec effect
    (c) parallax
    (d) direction finding
    (e) angular error

26. Suppose a ray of light, traveling at first through the air, strikes a flat pane of crown glass having uniform thickness at an angle of 30° relative to the normal. The index of refraction of the glass is 1.33. The ray goes through the glass and emerges into the air again. At what angle relative to the normal does the ray emerge?
    (a) 30°
    (b) 22°
    (c) 42°
    (d) The ray does not emerge, but is totally reflected within the glass
    (e) More information is needed to answer this question

27. Suppose a ray of light, traveling at first through the air, strikes a flat pane of flint glass having uniform thickness at an angle of 30° relative to the normal. The index of refraction of the glass is 1.52. The ray goes through the glass and emerges into the air again. At what angle relative to the normal does the ray emerge?
    (a) 30°
    (b) 19°
    (c) 50°
    (d) The ray does not emerge, but is totally reflected within the glass
    (e) More information is needed to answer this question

28. The upper equation in Fig. Test 2-3 expresses the spherical law of sines. This equation is useful if
    (a) we know the angular lengths of two of the sides of the spherical triangle QRS and the measure of the spherical angle between them, and we want to find the angular length of the third side
    (b) we know the measures of all three spherical angles and the angular length of one of the sides of the spherical triangle QRS, and we want to find the angular lengths of the other two sides
    (c) we know the angular lengths of all three sides of the spherical triangle QRS, and we want to find the radius of the sphere in linear units

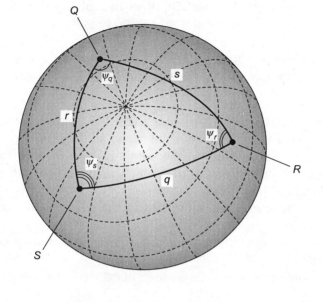

$$\frac{\sin q}{\sin \psi_q} = \frac{\sin r}{\sin \psi_r} = \frac{\sin s}{\sin \psi_s}$$

$$\cos s = \cos q \ \cos r + \sin q \ \sin r \ \cos \psi_s$$

Fig. Test 2-3. Illustration for Questions 28, 29, and 30 in the test for Part Two.

   (d) we know the angular length of one of the sides of the spherical triangle QRS, and we want to find the angular lengths of the other two sides

   (e) we know the measure of one of the spherical angles of the spherical triangle QRS, and we want to find the measures of the other two spherical angles

29. The lower equation in Fig. Test 2-3 expresses the spherical law of cosines. This equation is useful if

   (a) we know the angular lengths of two of the sides of the spherical triangle QRS and the measure of the spherical angle between them, and we want to find the angular length of the third side

   (b) we know the measures of all three spherical angles and the angular length of one of the sides of the spherical triangle QRS, and we want to find the angular lengths of the other two sides

   (c) we know the angular lengths of all three sides of the spherical triangle QRS, and we want to find the radius of the sphere in linear units

   (d) we know the angular length of one of the sides of the spherical triangle QRS, and we want to find the angular lengths of the other two sides

   (e) we know the measure of one of the spherical angles of the spherical triangle QRS, and we want to find the measures of the other two spherical angles

30. The two equations shown in Fig. Test 2-3 approach the laws of sines and cosines for triangles in a flat plane as the points Q, R, and S
   (a) become farther and farther apart relative to the size of the sphere
   (b) become more and more nearly the vertices of an equilateral spherical triangle
   (c) become more and more nearly the vertices of a right spherical triangle
   (d) become more and more nearly the vertices of an isosceles spherical triangle
   (e) become closer and closer together relative to the size of the sphere

31. Suppose a certain angle is stated as being equal to $10°$, plus or minus a measurement error of up to 1.00 minute of arc. What is this error figure, expressed as a percentage?
   (a) ±10.0%
   (b) ±1.67%
   (c) ±1.00%
   (d) ±0.167%
   (e) It is impossible to tell without more information

32. The absolute accuracy (in fixed units such as meters) with which the distance to an object can be measured using parallax depends on all of the following factors except:
   (a) the distance to the object
   (b) the length of the observation base line
   (c) the size of the object
   (d) the precision of the angle-measuring equipment
   (e) the distance between the two observation points

33. Imagine a sphere of shiny metal. Imagine a light source is so distant that the rays can be considered parallel. When reflected from the sphere, the light rays
   (a) diverge
   (b) converge
   (c) remain parallel

    (d) come to a focus

    (e) none of the above

34. The sum of the measures of the interior angles of a *spherical quadri-lateral* (a four-sided polygon on the surface of a sphere, all of whose sides are geodesic arcs) is always greater than

    (a) 360°

    (b) 540°

    (c) 630°

    (d) 720°

    (e) 810°

35. Imagine a spherical triangle with vertices Q, R, and S. Point Q is at the south geographic pole. Point R is on the equator at 30° east longitude. Point S is on the equator at 20° west longitude. What is the measure of $\angle_{sph} RQS$?

    (a) 20°

    (b) 30°

    (c) 50°

    (d) 90°

    (e) This question cannot be answered without more information

36. Imagine a spherical triangle with vertices Q, R, and S. Point Q is at the south geographic pole. Point R is on the equator at 30° east longitude. Point S is on the equator at 20° west longitude. What is the sum of the measures of the interior spherical angles of $\triangle_{sph} QRS$?

    (a) 180°

    (b) 200°

    (c) 210°

    (d) 230°

    (e) This question cannot be answered without more information

37. When a light ray passes through a boundary from a medium having an index of refraction r into a medium having an index of refraction s, the critical angle, θ_c, is given by the formula:

$$\theta_c = \arcsin\ (s/r)$$

What does this formula tell us about rays striking a boundary where $r = s$?

    (a) All incident rays pass through

    (b) No incident rays pass through

    (c) All incident rays pass through, except those striking normal to the boundary

     (d) No incident rays pass through, except those striking normal to the boundary

     (e) This formula tells us nothing at all if $r = s$

38. Imagine a spherical quadrilateral with vertices P, Q, R, and S. Point P is at the north geographic pole. Point Q is on the equator at 30° east longitude. Point R is at the south geographic pole. Point S is on the equator at 20° west longitude. What is the sum of the measures of the interior spherical angles of this spherical quadrilateral?

     (a) 360°

     (b) 410°

     (c) 460°

     (d) 540°

     (e) This question cannot be answered without more information

39. Which of the following must be a great circle on the surface of the earth?

     (a) Any circle on the surface that is centered at one of the poles

     (b) Any circle on the surface that is centered at a point on the Greenwich meridian

     (c) Any circle on the surface that is centered at a point on the equator

     (d) Any circle on the surface whose center coincides with the center of the earth

     (e) Any circle on the surface that passes through one of the poles

40. Suppose we observe a target on radar, and we see that its range is increasing while its azimuth (or bearing) is not changing. From this, we can conclude that

     (a) the target is heading north

     (b) the target is heading south

     (c) the target is heading east

     (d) the target is heading west

     (e) none of the above

41. What is the value of $2^3 \times 5 + 2 \times 3$?

     (a) 16

     (b) 46

     (c) 126

     (d) 168

     (e) This expression is ambiguous

42. Refer to the wave vector diagram of Fig. Test 2-4. Assume waves X and Y have identical frequency, and that the radial scale is linear. Which of the following is apparent?
    (a) Waves X and Y differ in peak amplitude by 50°
    (b) Waves X and Y are in phase opposition
    (c) Wave X lags wave Y by 50° of phase
    (d) Wave X leads wave Y by 50° of phase
    (e) The phase of wave Y is 4/5 of the phase of wave X

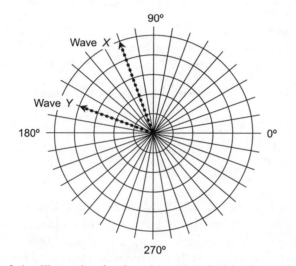

Fig. Test 2-4. Illustration for Questions 42 and 43 in the test for Part Two.

43. Refer to the wave vector diagram of Fig. Test 2-4. Assume waves X and Y have identical frequency, and that the radial scale is linear. Which of the following is apparent?
    (a) The peak amplitude of wave Y is 4/5 of the peak amplitude of wave X
    (b) The peak amplitude of wave X is 110°, and the peak amplitude of wave Y is 160°
    (c) Wave X travels in a different direction from wave Y
    (d) The dot product of X and Y is equal to 0
    (e) The cross product of X and Y is the zero vector

44. Imagine two alternating-current, sinusoidal waves X and Y that have the same frequency. Suppose wave X leads wave Y by 2π radians of phase. The more common way of saying this is
    (a) wave X leads wave Y by 2π radians
    (b) wave Y leads wave X by 2π radians

(c) wave Y lags wave X by $\pi/2$ radians

(d) the two waves are in phase coincidence

(e) nothing! The described situation is impossible

45. Suppose a surveyor measures the distance to an object using triangulation. As the length of the base line increases, assuming all other factors remain constant, the absolute accuracy (as expressed in terms of the maximum possible error in meters) of the distance measurement

  (a) improves

  (b) stays the same

  (c) gets worse

  (d) depends on factors not mentioned here

  (e) cannot be determined

46. The expression 3.457e−5 is another way of writing

  (a) 3457

  (b) 3.457

  (c) 3.457×10^5

  (d) 3.457×10^{-5}

  (e) none of the above

47. Refer to Fig. Test 2-5. A celestial object, which lies in the plane of the earth's orbit around the sun, is observed at two intervals three months apart, as shown. The angle θ is measured as $1.000°$. Recall that an astronomical unit (AU) is the mean distance of the earth from the sun. The distance to the celestial object, accurate to three significant figures, is

  (a) 1.00 AU

  (b) 57.3 AU

  (c) 100 AU

  (d) 360 AU

  (e) cannot be determined without more information

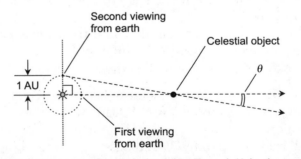

Fig. Test 2-5. Illustration for Questions 47, 48, and 49 in the test for Part Two.

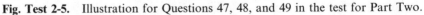

48. Suppose a celestial object is observed as in Fig. Test 2-5, and the angle θ for it is determined to be precisely $0°\ 00'\ 01''$. The distance to this object, accurate to four significant figures, is
    (a) 1000 AU
    (b) 100.0 AU
    (c) 10.00 AU
    (d) 1.000 AU
    (e) none of the above

49. Consider the scenario of Fig. Test 2-5 in general, for celestial objects that are many AU away from the earth. The size of the angle θ varies
    (a) in direct proportion to the square of the distance to an object
    (b) in direct proportion to the distance to an object
    (c) inversely as to the distance to the object
    (d) inversely as to the square of the distance to the object
    (e) none of the above

50. Suppose the measure of a certain angle is stated as $(4.66 \times 10^6)°$. From this, we can surmise that
    (a) the angle represents a tiny fraction of one revolution
    (b) the angle represents many revolutions
    (c) the sine of the angle is greater than 1
    (d) the sine of the angle is less than -1
    (e) the expression contains a typo, because angles cannot be expressed in power-of-10 notation

Final Exam

Do not refer to the text when taking this test. You may draw diagrams or use a calculator if necessary. A good score is at least 75 correct. Answers are in the back of the book. It's best to have a friend check your score the first time, so you won't memorize the answers if you want to take the test again.

1. Suppose $y = \csc x$. The domain of this function includes all real numbers except
   (a) integral multiples of $\pi/4$ rad
   (b) integral multiples of $\pi/2$ rad
   (c) integral multiples of π rad
   (d) integral multiples of 2π rad
   (e) integral multiples of 4π rad

2. The expression $\sin^{-1}(x)$ is equivalent to the expression
   (a) $1/(\sin x)$
   (b) $\arcsin x$
   (c) $\sin x - 1$
   (d) $\sin (x - 1)$
   (e) $-\sin x$

3. In Fig. Exam-1, which of the graphs represent functions of x?
   (a) L only

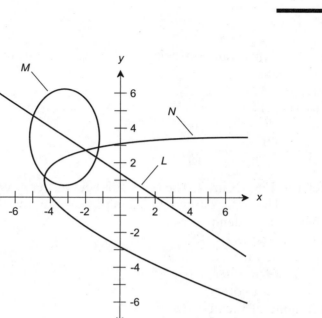

Fig. Exam-1. Illustration for Questions 3, 4, and 5 in the final exam.

(b) L and N
(c) M and N
(d) M only
(e) none of them

4. The coordinate scheme in Fig. Exam-1 is an example of
   (a) a polar system
   (b) an equilateral system
   (c) a logarithmic system
   (d) a rectangular system
   (e) Euclidean three-space

5. In a coordinate system such as that shown in Fig. Exam-1, how far is the point $(3,-4)$ from the origin?
   (a) 3 units
   (b) 4 units
   (c) -3 units
   (d) -4 units
   (e) none of the above

6. The cosine of the negative of an angle is equal to the cosine of the angle. The following formula holds for any angle θ:

$$\cos -\theta = \cos \theta$$

Based on this trigonometric identity, we can conclude that cos 320° is the same as

(a) −cos 320°
(b) −cos 40°
(c) cos 40°
(d) sin 40°
(e) −sin 40°

7. Refer to Fig. Exam-2. The bold, solid curves represent the function $y = \coth x$. The bold, dashed curves represent the inverse of this function, which can be denoted as

(a) $y = (\coth x)^{-1}$
(b) $y = \coth^{-1} x$
(c) $y = 1/(\coth x)$
(d) $x = \operatorname{arc} \coth y$
(e) none of the above

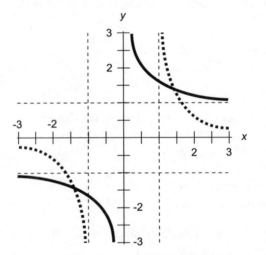

Fig. Exam-2. Illustration for Questions 7, 8, and 9 in the final exam.

8. Refer to Fig. Exam-2. Which of the following quantities (a), (b), or (c) is in the domain of the function shown by the bold, dashed curves?

(a) $x = -1$
(b) $x = 0$
(c) $x = 1$
(d) All of the above quantities (a), (b), and (c) are in the domain
(e) None of the above quantities (a), (b), or (c) are in the domain

9. Refer to Fig. Exam-2. Which of the following quantities (a), (b), or (c) lies outside the domain of the function shown by the bold, solid curves?
   (a) $x = -1$
   (b) $x = 0$
   (c) $x = 1$
   (d) All of the above quantities (a), (b), and (c) are outside the domain
   (e) None of the above quantities (a), (b), or (c) are outside the domain

10. The arctangent function, $y = \arctan x$, is defined for
   (a) all values of x
   (b) $x > 0$ only
   (c) $x < 0$ only
   (d) $-2\pi < x < 2\pi$ only
   (e) $-1 < x < 1$ only

11. Triangulation using parallax involves the measurement of distance by observing an object from
   (a) a single reference point
   (b) two reference points that lie on a ray pointing in the direction of the distant object
   (c) two reference points that lie on a line perpendicular to a ray pointing in the direction of the distant object
   (d) three reference points that lie on a ray pointing in the direction of the distant object
   (e) three reference points that lie on a line perpendicular to a ray pointing in the direction of the distant object

12. Suppose you know the lengths of two sides p and q of a triangle, and the measure of the angle θ_r between them. Then the length of the third side r is:

$$r = (p^2 + q^2 - 2pq \, \cos \, \theta_r)^{1/2}$$

Recall this as the law of cosines. Knowing this, suppose you are at the intersection of two roads. One road runs exactly east/west, and the other runs exactly northeast/southwest. You see a car on one road 500 meters to your southwest, and a car on the other road 700 meters to your east. How far from each other are the cars, as measured along a straight line?
   (a) 900 meters
   (b) 1.11 kilometers
   (c) 1.20 kilometers

(d) 1.41 kilometers

(e) It is impossible to calculate this without more information

13. Refer to Fig. Exam-3. How many orders of magnitude does the horizontal scale encompass?

(a) 3

(b) 7

(c) 10

(d) 21

(e) Infinitely many

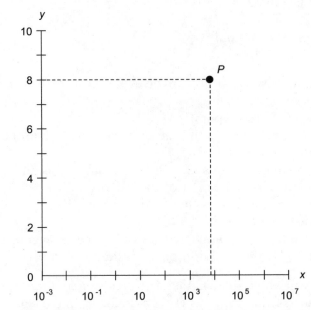

Fig. Exam-3. Illustration for Questions 13, 14, and 15 in the final exam.

14. Refer to Fig. Exam-3. How many orders of magnitude does the vertical scale encompass?

(a) 0

(b) 1

(c) 10

(d) 100

(e) Infinitely many

15. In Fig. Exam-3, by how many orders of magnitude (approximately) do the x and y coordinates of point P differ?

(a) 1

(b) 3

(c) 8

(d) Infinitely many

(e) More information is needed to tell

16. In exponential terms, the hyperbolic sine (sinh) function of a variable x can be expressed in the form:

$$\sinh x = (e^x - e^{-x})/2$$

According to this formula, as x becomes larger and larger negatively without limit, what happens to the value of sinh x?

(a) It becomes larger and larger positively, without limit

(b) It approaches 0 from the positive direction

(c) It becomes larger and larger negatively, without limit

(d) It approaches 0 from the negative direction

(e) It alternates endlessly between negative and positive values

17. In scientific notation, an exponent takes the form of

(a) a subscript

(b) an italicized numeral

(c) a superscript

(d) a boldface quantity

(e) an underlined quantity

18. Suppose we are confronted with the following combination of products and a quotient after having conducted a scientific experiment involving measurements:

$$(3.55 \times 290.992)/(64.24 \times 796.66)$$

How many significant figures can we claim when we calculate the result?

(a) 2

(b) 3

(c) 4

(d) 5

(e) 6

19. If the frequency of a wave is 1000 Hz, then the period of the wave is

(a) 0.001000 second

(b) 0.00628 second

(c) 0.360 second

(d) 1.000 second

(e) impossible to determine without more information

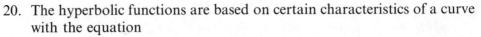
20. The hyperbolic functions are based on certain characteristics of a curve with the equation
    (a) $x + y = 1$
    (b) $x - y = 1$
    (c) $x^2 + y^2 = 1$
    (d) $x^2 - y^2 = 1$
    (e) $y = x^2 + 2x + 1$

21. Suppose the coordinates of a point in the mathematician's polar plane are specified as $(\theta, r) = (-\pi/4, -2)$. This is equivalent to the coordinates
    (a) $(\pi/4, 2)$
    (b) $(3\pi/4, 2)$
    (c) $(5\pi/4, 2)$
    (d) $(7\pi/4, 2)$
    (e) none of the above

22. Figure Exam-4 illustrates an example of distance measurement by means of
    (a) angular deduction
    (b) triangulation
    (c) the law of sines
    (d) stadimetry
    (e) parallax comparison

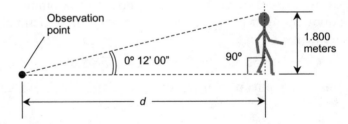

Fig. Exam-4. Illustration for Questions 22, 23, and 24 in the final exam.

23. Approximately what is the distance d in the scenario of Fig. Exam-4?
    (a) 8.47 meters
    (b) 516 meters
    (c) 859 meters
    (d) 30.9 kilometers
    (e) It is impossible to determine without more information

24. In the scenario of Fig. Exam-4, suppose the distance d doubles, while the human's height and orientation do not change. Approximately

what will be the angular height (or diameter) of the human, as seen from the same point of observation?
- (a) 0° 48′ 00″
- (b) 0° 24′ 00″
- (c) 0° 12′ 00″
- (d) 0° 06′ 00″
- (e) 0° 03′ 00″

25. Snell's law is a principle that involves
- (a) the behavior of refracted light rays
- (b) hyperbolic functions
- (c) cylindrical-to-spherical coordinate conversion
- (d) Cartesian-to-polar coordinate conversion
- (e) wave amplitude versus frequency

26. Fill in the blank to make the following statement the most correct and precise: "In optics, the angle of incidence is usually expressed with respect to a line _____ the surface at the point where reflection takes place."
- (a) parallel to
- (b) passing through
- (c) normal to
- (d) tangent to
- (e) that does not intersect

27. Suppose a prism is made out of glass that has an index of refraction of 1.45 at all visible wavelengths. If this prism is placed in a liquid that also has an index of refraction of 1.45 at all visible wavelengths, then
- (a) rays of light encountering the prism will behave just as they do when the prism is surrounded by any other transparent substance
- (b) rays of light encountering the prism will all be reflected back into the liquid
- (c) rays of light encountering the prism will pass straight through it as if it were not there
- (d) some of the light entering the prism will be trapped inside by total internal reflection
- (e) all of the light entering the prism will be trapped inside by total internal reflection

28. On a radar display, a target appears at azimuth 280°. This is
- (a) 10° east of south
- (b) 10° west of south
- (c) 10° south of west

(d) $10°$ west of north

(e) none of the above

29. Suppose a pair of tiny, dim stars in mutual orbit, never before seen because we didn't have powerful enough telescopes, is discovered at a distance of 1 parsec from our Solar System. When the stars are at their maximum angular separation as observed by our telescopes, they are $\frac{1}{2}$ second of arc apart. What is the actual distance between these stars, in astronomical units (AU), when we see them at their maximum angular separation? Remember that an astronomical unit is defined as the mean distance of the earth from the sun.

(a) This question cannot be answered without more information

(b) $\frac{1}{4}$ AU

(c) $\frac{1}{2}$ AU

(d) 1 AU

(e) 2 AU

30. Suppose two vectors are oriented at a $60°$ angle relative to each other. The length of vector **a** is exactly 6 units, and the length of vector **b** is exactly 2 units. What is the dot product **a** · **b**, accurate to three significant figures?

(a) 0.00

(b) 6.00

(c) 10.4

(d) 12.0

(e) More information is necessary to answer this question

31. On a sunny day, your shadow is half as great as your height when the sun is

(a) $15°$ from the zenith

(b) $45°$ from the zenith

(c) $60°$ from the zenith

(d) $75°$ from the zenith

(e) none of the above

32. When a light ray passes through a boundary from a medium having an index of refraction r into a medium having an index of refraction s, the critical angle, θ_c, is given by the formula:

$$\theta_c = \arcsin(s/r)$$

What does this formula tell us about rays striking a boundary where $r = s/2$?

(a) Only those rays striking at an angle of incidence less than 60° pass through
(b) Only those rays striking at an angle of incidence greater than 60° pass through
(c) Only those rays striking at an angle of incidence less than 30° pass through
(d) Only those rays striking at an angle of incidence greater than 30° pass through
(e) The critical angle is not defined if $r = s/2$

33. A geodesic that circumnavigates a sphere is also called
    (a) a spherical circle
    (b) a parallel
    (c) a meridian
    (d) a great circle
    (e) a spherical arc

34. The sum of the measures of the interior angles of a *spherical pentagon* (a five-sided polygon on the surface of a sphere, all of whose sides are geodesic arcs) is always greater than
    (a) 540°
    (b) 630°
    (c) 720°
    (d) 810°
    (e) 900°

35. What is the shortest possible height for a flat wall mirror that allows a man 180 centimeters tall to see his full reflection?
    (a) 180 centimeters
    (b) 135 centimeters
    (c) 127 centimeters
    (d) 90 centimeters
    (e) It depends on the distance between the man and the mirror

36. Imagine four distinct points on the earth's surface. Two of the points are on the Greenwich meridian (longitude 0°) and two of them are at longitude 180°. Suppose each adjacent pair of points is connected by an arc representing the shortest possible path over the earth's surface. What is the sum of the measures of the interior angles of the resulting spherical quadrilateral?
    (a) 360°
    (b) 540°
    (c) 720°

(d) More information is needed to answer this question

(e) It cannot be defined

37. Imagine four distinct points on the earth's surface, all of which lie on the equator. Suppose each adjacent pair of points is connected by an arc representing the shortest possible path over the earth's surface. What is peculiar about the resulting spherical quadrilateral?

(a) The inside of the quadrilateral can just as well be called the outside, and the outside can just as well be called the inside

(b) All four sides have the same angular length, but all four interior spherical angles have different measures

(c) No two sides can have the same angular length

(d) The interior area of the quadrilateral is greater than the surface area of the earth

(e) The interior area of the quadrilateral cannot be calculated

38. The cotangent of an angle is equal to

(a) the sine divided by the cosine, provided the cosine is not equal to zero

(b) the cosine divided by the sine, provided the sine is not equal to zero

(c) 1 minus the tangent

(d) $90°$ minus the tangent

(e) the sum of the squares of the sine and the cosine

39. The hyperbolic secant of a quantity x, symbolized sech x, can be defined according to the following formula:

$$\text{sech } x = 2/(e^x + e^{-x})$$

For which, if any, of the following values of x is this function undefined?

(a) $-1 < x < 1$

(b) $0 < x < 1$

(c) $-1 < x < 0$

(d) $x < 0$

(e) None of the above; the function is defined for all real-number values of x

40. Written in scientific notation, the number 255,308 is

(a) 255308

(b) 0.255308×10^5

(c) 2.55308×10^5

(d) 0.255308×10^{-5}

(e) 2.55308×10^{-5}

41. Figure Exam-5 shows the path of a light ray R, which becomes ray S as it crosses a flat boundary B between media having two different indexes of refraction r and s. Suppose that line N is normal to plane B. Also suppose that line N, ray R, and ray S all intersect plane B at point P. If $\theta = 55°$ and $\phi = 30°$, we can conclude that
   (a) $r > s$
   (b) $r = s$
   (c) $r < s$
   (d) the illustrated situation is impossible
   (e) rays R and S cannot lie in the same plane

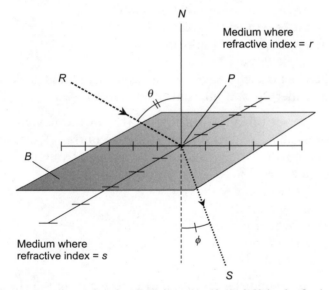

Fig. Exam-5. Illustration for Questions 41, 42, and 53 in the final exam.

42. Imagine a light ray R, which becomes ray S as it crosses a flat boundary B between media having two different indexes of refraction r and s, as shown in Fig. Exam-5. Suppose that line N is normal to plane B. Also suppose that line N, ray R, and ray S all intersect plane B at point P. We are given the following equation relating various parameters in this situation:

$$s \, \sin \, \phi = r \, \sin \, \theta$$

Suppose we are told, in addition to all of the above information, that $\theta = 55° \, 00'$, $\phi = 30° \, 00'$, and $r = 1.000$. From this, we can determine that
   (a) $s = 1.638$
   (b) $s = 0.410$

(c) $s = 1.833$
(d) $s = 1.000$
(e) none of the above

43. Imagine a light ray R, which encounters a flat boundary B between media having two different indexes of refraction r and s, as shown in Fig. Exam-5. Suppose that line N is normal to plane B. Also suppose that line N and ray R intersect plane B at point P. Suppose we are told that $r < s$. What can we conclude about the angle of incidence θ at which ray R undergoes total internal reflection at the boundary plane B?
    (a) The angle θ must be greater than $0°$
    (b) The angle θ must be greater than $45°$
    (c) The angle θ must be less than $90°$
    (d) The angle θ must be less than $45°$
    (e) There is no such angle θ, because no ray R that strikes B as shown can undergo total internal reflection if $r < s$

44. Suppose we set off on a bearing of $315°$ in the navigator's polar coordinate system. We stay on a straight course. If the starting point is considered the origin, what is the graph of our path in Cartesian coordinates?
    (a) $y = -x$, where $x \leq 0$
    (b) $y = 0$, where $x \geq 0$
    (c) $x = 0$, where $y \geq 0$
    (d) $y = -x$, where $x \geq 0$
    (e) None of the above

45. What is the angular length of an arc representing the shortest possible distance over the earth's surface connecting the south geographic pole with the equator?
    (a) $0°$
    (b) $45°$
    (c) $90°$
    (d) $135°$
    (e) It is impossible to answer this without knowing the longitude of the point where the arc intersects the equator

46. Minneapolis, Minnesota is at latitude $+45°$. What is the angular length of an arc representing the shortest possible distance over the earth's surface connecting Minneapolis with the south geographic pole?
    (a) $0°$
    (b) $45°$

(c) 90°

(d) 135°

(e) It is impossible to answer this without knowing the longitude of Minneapolis

47. When a light ray passes through a boundary from a medium having an index of refraction r into a medium having an index of refraction s, the critical angle, θ_c, is given by the formula:

$$\theta_c = \arcsin (s/r)$$

Suppose $\theta_c = 1$ rad, and $s = 1.225$. What is r?

(a) 0.687

(b) 1.031

(c) 1.456

(d) We need more information to answer this question

(e) It is undefined; such a medium cannot exist

48. The equal-angle axes in the mathematician's polar coordinate system are

(a) rays

(b) spirals

(c) circles

(d) ellipses

(e) hyperbolas

49. The dot product of two vectors that point in opposite directions is

(a) a vector with zero magnitude

(b) a negative real number

(c) a positive real number

(d) a vector perpendicular to the line defined by the two original vectors

(e) a vector parallel to the line defined by the two original vectors

50. The cross product of two vectors that point in opposite directions is

(a) a vector with zero magnitude

(b) a negative real number

(c) a positive real number

(d) a vector perpendicular to the line defined by the two original vectors

(e) a vector parallel to the line defined by the two original vectors

51. What is the phase difference, in radians, between the two waves defined by the following functions:

$$y = -2 \sin x$$
$$y = 3 \sin x$$

(a) 0
(b) $\pi/4$
(c) $\pi/2$
(d) π
(e) It is undefined, because the two waves do not have the same frequency

52. What is the phase difference, in radians, between the two waves defined by the following functions:

$$y = -3 \sin x$$
$$y = 5 \cos x$$

(a) 0
(b) $\pi/4$
(c) $\pi/2$
(d) π
(e) It is undefined, because the two waves do not have the same frequency

53. What is the phase difference, in radians, between the two waves defined by the following functions:

$$y = -4 \cos x$$
$$y = -6 \cos x$$

(a) 0
(b) $\pi/4$
(c) $\pi/2$
(d) π
(e) It is undefined, because the two waves do not have the same frequency

54. Suppose there are two sine waves *X* and *Y*. The frequency of wave *X* is 350 Hz, and the frequency of wave *Y* is 360 Hz. From this, we know that
(a) wave *X* leads wave *Y* by 10° of phase
(b) wave *X* lags wave *Y* by 10° of phase
(c) the amplitudes of the waves differ by 10 Hz
(d) the phases of the waves differ by 10 Hz
(e) none of the above

55. Suppose a distant celestial object is observed, and its angular diameter is said to be $0° \, 0' \, 0.5000'' \pm 10\%$. This indicates that the angular diameter is somewhere between
    (a) $0° \, 0' \, 0.4000''$ and $0° \, 0' \, 0.6000''$
    (b) $0° \, 0' \, 0.4500''$ and $0° \, 0' \, 0.5500''$
    (c) $0° \, 0' \, 0.4900''$ and $0° \, 0' \, 0.5100''$
    (d) $0° \, 0' \, 0.4950''$ and $0° \, 0' \, 0.5050''$
    (e) $0° \, 0' \, 0.4995''$ and $0° \, 0' \, 0.5005''$

56. Suppose there are two sine waves X and Y having identical frequency. Suppose that in a vector diagram, the vector for wave X is $80°$ clockwise from the vector representing wave Y. This means that
    (a) wave X leads wave Y by $80°$
    (b) wave X leads wave Y by $110°$
    (c) wave X lags wave Y by $80°$
    (d) wave X lags wave Y by $110°$
    (e) none of the above

57. In navigator's polar coordinates, it is important to specify whether $0°$ refers to magnetic north or geographic north. At a given location on the earth, the difference, as measured in degrees of the compass, between magnetic north and geographic north is called
    (a) azimuth imperfection
    (b) polar deviation
    (c) equatorial inclination
    (d) right ascension
    (e) declination

58. Refer to Fig. Exam-6. Given that the size of the sphere is constant, the length of arc QR approaches the length of line segment QR as
    (a) points Q and R become closer and closer to point P
    (b) points Q and R become closer and closer to each other
    (c) points Q and R become farther and farther from point P
    (d) points Q and R become farther and farther from each other
    (e) none of the above

59. Refer to Fig. Exam-6. What is the greatest possible length of line segment QR?
    (a) Half the circumference of the sphere
    (b) The circumference of the sphere
    (c) Twice the radius of the sphere
    (d) The radius of the sphere
    (e) None of the above

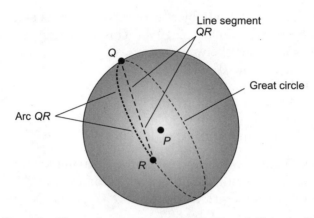

Fig. Exam-6. Illustration for Questions 58, 59, and 60 in the final exam.

60. Suppose, in the scenario shown by Fig. Exam-6, point Q remains stationary while point R revolves around the great circle, causing the length of arc QR to increase without limit (we allow the arc to represent more than one complete trip around the sphere). As this happens, the length of line segment QR
    (a) oscillates between zero and a certain maximum, over and over
    (b) increases without limit
    (c) reaches a certain maximum and then stays there
    (d) becomes impossible to define
    (e) none of the above

61. Suppose that the measure of angle θ in Fig. Exam-7 is $27°$. Then the measure of $\angle QRP$ is
    (a) $18°$
    (b) $27°$
    (c) $63°$
    (d) $153°$
    (e) impossible to determine without more information

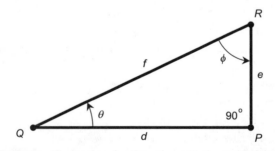

Fig. Exam-7. Illustration for Questions 61 through 64 in the final exam.

62. In Fig. Exam-7, the ratio e/f represents
    (a) $\cos \phi$
    (b) $\cos \theta$
    (c) $\tan \phi$
    (d) $\tan \theta$
    (e) $\sec \theta$

63. In Fig. Exam-7, $\csc \phi$ is represented by the ratio
    (a) d/f
    (b) d/e
    (c) e/f
    (d) f/e
    (e) f/d

64. In Fig. Exam-7, which of the following is true?
    (a) $\sin^2 \theta + \cos^2 \theta = 1$
    (b) $\sin^2 \theta + \cos^2 \phi = 1$
    (c) $\sin^2 \theta + \cos^2 \phi = 0$
    (d) $\theta - \phi = \pi/2$ rad
    (e) None of the above

65. What is the value of arctan (-1) in radians? Consider the range of the arctangent function to be limited to values between, but not including, $-\pi/2$ rad and $\pi/2$ rad. Do not use a calculator to determine the answer.
    (a) $-\pi/3$
    (b) $-\pi/4$
    (c) 0
    (d) $\pi/4$
    (e) $\pi/3$

66. Suppose a target is detected 10 kilometers east and 13 kilometers north of our position. The azimuth of this target is approximately
    (a) $38°$
    (b) $52°$
    (c) $128°$
    (d) $142°$
    (e) impossible to calculate without more information

67. Suppose a target is detected 20 kilometers west and 48 kilometers south of our position. The distance to this target is approximately
    (a) 68 kilometers
    (b) 60 kilometers
    (c) 56 kilometers

(d) 52 kilometers

(e) impossible to calculate without more information

68. Suppose an airborne target appears on a navigator's-polar-coordinate radar display at azimuth 270°. The target flies on a heading directly north, and continues on that heading. As we watch the target on the radar display

(a) its azimuth and range both increase

(b) its azimuth increases and its range decreases

(c) its azimuth decreases and its range increases

(d) its azimuth and range both decrease

(e) its azimuth and range both remain constant

69. In 5/8 of an alternating-current wave cycle, there are

(a) 45° of phase

(b) 90° of phase

(c) 135° of phase

(d) 180° of phase

(e) 225° of phase

70. In cylindrical coordinates, the position of a point is specified by

(a) two angles and a distance

(b) two distances and an angle

(c) three distances

(d) three angles

(e) none of the above

71. The expression 3 cos 60° + 2 tan 45°/sin 30° is

(a) ambiguous

(b) equal to 5.5

(c) equal to 7

(d) equal to 27

(e) undefined

72. The sine of an angle can be at most equal to

(a) 1

(b) π

(c) 2π

(d) 180°

(e) anything! There is no limit to how large the sine of an angle can be

73. Suppose you see a balloon hovering in the sky over a calm ocean. You are told that it is 10 kilometers north of your position, 10 kilometers east of your position, and 10 kilometers above the surface of the ocean.

This information is an example of the position of the balloon expressed in a form of
(a) Cartesian coordinates
(b) cylindrical coordinates
(c) spherical coordinates
(d) celestial coordinates
(e) none of the above

74. In Fig. Exam-8, the frequencies of waves X and Y appear to
(a) differ by a factor of about 2
(b) be about the same
(c) differ by about 180°
(d) differ by about $\pi/2$ radians
(e) none of the above

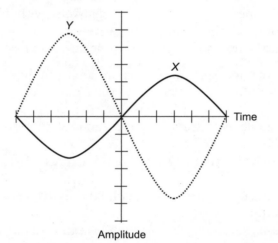

Fig. Exam-8. Illustration for Questions 74, 75, and 76 in the final exam.

75. In Fig. Exam-8, the phases of waves X and Y appear to
(a) differ by a factor of about 2
(b) be about the same
(c) differ by about 180°
(d) differ by about $\pi/2$ radians
(e) none of the above

76. In Fig. Exam-8, the amplitudes of waves X and Y appear to
(a) differ by a factor of about 2
(b) be about the same
(c) differ by about 180°

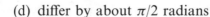
(d) differ by about $\pi/2$ radians

(e) none of the above

77. Which, if any, of the following expressions (a), (b), (c), or (d) is undefined?

(a) $\sin 0°$

(b) $\sin 90°$

(c) $\cos \pi$ rad

(d) $\cos 2\pi$ rad

(e) All of the above expressions are defined

78. As $x \to 0^+$ (that is, x approaches 0 from the positive direction), what happens to the value of $\ln x$ (the natural logarithm of x)?

(a) It becomes larger and larger positively, without limit

(b) It approaches 0 from the positive direction

(c) It becomes larger and larger negatively, without limit

(d) It approaches 0 from the negative direction

(e) It alternates endlessly between negative and positive values

79. Suppose the measure of a certain angle in mathematician's polar coordinates is stated as -9.8988×10^{-75} rad. From this, we can surmise that

(a) the angle is extremely large, and is expressed in a clockwise direction

(b) the angle is extremely large, and is expressed in a counterclockwise direction

(c) the angle is extremely small, and is expressed in a clockwise direction

(d) the angle is extremely small, and is expressed in a counterclockwise direction

(e) the expression contains a typo, because angles cannot be negative

80. The hyperbolic cosine of a quantity x, symbolized cosh x, can be defined according to the following formula:

$$\cosh x = (e^x + e^{-x})/2$$

Based on this, what is the value of cosh 0? You should not need a calculator to figure this out.

(a) 0

(b) 1

(c) 2

(d) -1

(e) -2

81. An abscissa is
    (a) a coordinate representing a variable
    (b) the shortest path between two points
    (c) a vector perpendicular to a specified plane
    (d) the origin of a coordinate system
    (e) the boundary of a coordinate system

82. Suppose you are standing at the north geographic pole. Suppose you
    fire two guns, call them A and B, simultaneously in horizontal direc-
    tions, gun A along the Prime Meridian ($0°$ longitude) and gun B along
    the meridian representing $+90°$ ($90°$ east longitude). Suppose the
    bullets from both guns travel at 5000 meters per second. Let **a**
    be the vector representing the velocity of the bullet from gun A; let **b**
    be the vector representing the velocity of the bullet from gun B. What is
    the direction of vector **a** × **b** the instant after the guns are fired?
    (a) $+45°$ ($45°$ east longitude)
    (b) Straight up
    (c) Straight down
    (d) Undefined, because the magnitude of **a** × **b** is zero
    (e) This question cannot be answered without more information

83. In Fig. Exam-9, θ_x, θ_y, and θ_z
    (a) represent variables in spherical coordinates
    (b) represent azimuth, elevation, and declination
    (c) are always expressed in a clockwise rotational sense

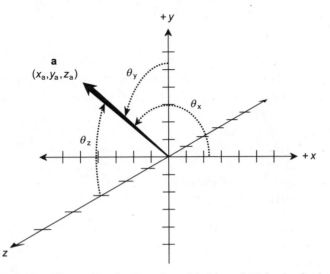

Fig. Exam-9. Illustration for Questions 83, 84, and 85 in the final exam.

(d) uniquely define the direction of vector **a**

(e) uniquely define the magnitude of vector **a**

84. Refer to Fig. Exam-9. Suppose the direction (or orientation) in which vector **a** points is exactly reversed. What happens to θ_x, θ_y, and θ_z?
    (a) Their measures all change by 180°
    (b) Their measures all remain the same
    (c) Their measures are all multiplied by -1
    (d) Their measures all increase by $\pi/2$ rad
    (e) It is impossible to say without more information

85. Refer to Fig. Exam-9. Suppose the values of x_a, y_a, and z_a are all doubled. What happens to θ_x, θ_y, and θ_z?
    (a) Their measures are all doubled
    (b) Their measures all remain the same
    (c) Their measures are all quadrupled
    (d) Their measures are all divided by 2
    (e) It is impossible to say without more information

86. Consider the circle represented by the equation $x^2 + y^2 = 9$ on the Cartesian plane. Imagine a ray running outward from the origin through a point on the circle where $x = y$. Consider the angle between the ray and the positive x axis, measured counterclockwise. The tangent of this angle is equal to
    (a) $x/3$
    (b) $y/3$
    (c) 1
    (d) 0
    (e) 3

87. Suppose a computer display has an aspect ratio of 4:3. This means that the width is 4/3 times the height. A diagonal line on this display is slanted at approximately
    (a) 30° relative to horizontal
    (b) 37° relative to horizontal
    (c) 45° relative to horizontal
    (d) 53° relative to horizontal
    (e) 60° relative to horizontal

88. Suppose a geometric object in the polar coordinate plane is represented by the equation $r = -3$. The object is
    (a) a circle
    (b) a hyperbola

(c) a parabola
(d) a straight line
(e) a spiral

89. Suppose a geometric object in the polar coordinate plane is represented by the equation $\theta = 3\pi/4$. The object is
(a) a circle
(b) a hyperbola
(c) a parabola
(d) a straight line
(e) a spiral

90. The hyperbolic functions are
(a) inverses of the circular functions
(b) negatives of the circular functions
(c) reciprocals of the circular functions
(d) identical with the circular functions
(e) none of the above

91. Refer to Fig. Exam-10. What are the coordinates of point P? Assume that the curves intersect there.
(a) $(-5\pi/4, 2^{1/2})$
(b) $(-5\pi/4, 2^{-1/2})$
(c) $(-7\pi/4, 2^{1/2})$
(d) $(-7\pi/4, 2^{-1/2})$
(e) They cannot be determined without more information

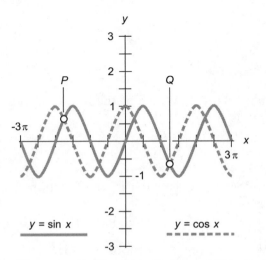

Fig. Exam-10. Illustration for Questions 91, 92, and 93 in the final exam.

92. Refer to Fig. Exam-10. What are the coordinates of point Q? Assume that the curves intersect there.
    (a) $(5\pi/4, -2^{1/2})$
    (b) $(5\pi/4, -2^{-1/2})$
    (c) $(7\pi/4, -2^{1/2})$
    (d) $(7\pi/4, -2^{-1/2})$
    (e) They cannot be determined without more information

93. Refer to Fig. Exam-10. By what extent is the cosine wave displaced along the x axis relative to the sine wave?
    (a) 180° negatively
    (b) 135° negatively
    (c) 90° negatively
    (d) 45° negatively
    (e) This question cannot be answered without more information

94. How many radians are there in an angle representing three-quarters of a circle?
    (a) 0.25π
    (b) 0.75π
    (c) π
    (d) 1.5π
    (e) This question is meaningless, because the radian is not a unit of angular measure

95. Which of the following functions has a graph that is not sinusoidal?
    (a) $f(x) = 3 \sin x$
    (b) $f(x) = -2 \cos 2x$
    (c) $f(x) = 4 \csc 4x$
    (d) $f(x) = 4 \cos (-3x)$
    (e) $f(x) = -\cos (\pi x)$

96. Suppose $f(x) = 3x + 1$. Which of the following statements (a), (b), (c), or (d), if any, is true?
    (a) $f^{-1}(x) = (x - 1)/3$
    (b) $f^{-1}(x) = x/3 + 1/3$
    (c) $f^{-1}(x) = -3x - 1$
    (d) $f^{-1}(x)$ does not exist; that is, the function $f(x) = 3x + 1$ has no inverse
    (e) None of the above statements (a), (b), (c), or (d) is true

97. Which of the following expressions is undefined?
    (a) $\csc 0°$

(b) sec 0°

(c) tan 45°

(d) sin 180°

(e) cot 135°

98. Refer to Fig. Exam-11. If the rectangular coordinates x_0 and y_0 of point P are both doubled, what happens to the value of r_0?
   (a) It increases by a factor of the square root of 2
   (b) It doubles
   (c) It quadruples
   (d) It does not change
   (e) This question cannot be answered without more information

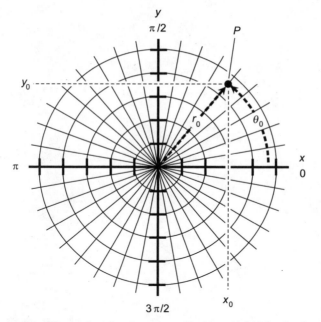

Fig. Exam-11. Illustration for Questions 98, 99, and 100 in the final exam.

99. Refer to Fig. Exam-11. If the rectangular coordinates x_0 and y_0 of point P are both doubled, what happens to the value of θ_0?
   (a) It increases by a factor of the square root of 2
   (b) It doubles
   (c) It is multiplied by -1
   (d) It does not change
   (e) It increases by π rad

100. Refer to Fig. Exam-11. If the rectangular coordinates x_0 and y_0 of point P are both multiplied by -1, what happens to the value of θ_0?
    (a) It increases by a factor of the square root of 2
    (b) It doubles
    (c) It is multiplied by -1
    (d) It does not change
    (e) It increases by π rad

Answers to Quiz, Test, and Exam Questions

CHAPTER 1

1. d	2. d	3. a	4. c	5. a
6. c	7. b	8. d	9. b	10. d

CHAPTER 2

1. b	2. a	3. d	4. a	5. c
6. d	7. a	8. d	9. c	10. d

CHAPTER 3

1. b	2. c	3. b	4. d	5. a
6. b	7. a	8. d	9. c	10. a

CHAPTER 4

1. a	2. d	3. d	4. a	5. d
6. a	7. b	8. c	9. b	10. a

CHAPTER 5

1. c	2. b	3. c	4. c	5. d
6. a	7. c	8. b	9. a	10. a

CHAPTER 6

1. c	2. c	3. a	4. b	5. a
6. d	7. d	8. c	9. a	10. b

TEST: PART ONE

1. d	2. d	3. e	4. d	5. a
6. e	7. c	8. d	9. b	10. b
11. d	12. e	13. b	14. c	15. a
16. e	17. a	18. d	19. b	20. c
21. a	22. b	23. c	24. a	25. b
26. a	27. e	28. e	29. d	30. b
31. c	32. e	33. e	34. b	35. d
36. c	37. a	38. b	39. e	40. b
41. a	42. e	43. c	44. a	45. c
46. d	47. a	48. b	49. e	50. d

CHAPTER 7

1. c	2. d	3. c	4. a	5. c
6. b	7. b	8. b	9. a	10. d

CHAPTER 8

1. c	2. b	3. d	4. c	5. c
6. c	7. b	8. b	9. a	10. a

CHAPTER 9

1. b	2. c	3. b	4. c	5. c
6. d	7. d	8. a	9. b	10. d

CHAPTER 10

1. d	2. a	3. c	4. b	5. d
6. b	7. d	8. b	9. b	10. a

CHAPTER 11

1. d	2. b	3. d	4. c	5. c
6. b	7. a	8. a	9. c	10. c

TEST: PART TWO

1. d	2. c	3. d	4. e	5. b
6. e	7. a	8. e	9. c	10. e
11. a	12. b	13. c	14. d	15. d
16. a	17. b	18. a	19. c	20. a
21. c	22. c	23. a	24. e	25. c
26. a	27. a	28. b	29. a	30. e
31. d	32. c	33. a	34. a	35. c
36. d	37. a	38. c	39. d	40. e
41. b	42. c	43. a	44. d	45. a
46. d	47. b	48. e	49. c	50. b

FINAL EXAM

1. c	2. b	3. a	4. d	5. e
6. c	7. b	8. e	9. b	10. a
11. c	12. b	13. c	14. e	15. b

Answers

16. c	17. c	18. b	19. a	20. d
21. b	22. d	23. b	24. d	25. a
26. c	27. c	28. e	29. c	30. b
31. e	32. e	33. d	34. a	35. d
36. c	37. a	38. b	39. e	40. c
41. c	42. a	43. e	44. a	45. c
46. d	47. c	48. a	49. b	50. a
51. d	52. c	53. a	54. e	55. b
56. c	57. e	58. b	59. c	60. a
61. c	62. a	63. e	64. a	65. b
66. a	67. d	68. a	69. e	70. b
71. b	72. a	73. a	74. b	75. c
76. a	77. e	78. c	79. c	80. b
81. a	82. b	83. d	84. a	85. b
86. c	87. b	88. a	89. d	90. e
91. d	92. b	93. c	94. d	95. c
96. a	97. a	98. b	99. d	100. e

Suggested Additional References

Books

Downing, Douglas, *Trigonometry the Easy Way*. Hauppauge, NY, Barron's
Educational Series, Inc., 2001.
Gibilisco, Stan, *Geometry Demystified*. New York, McGraw-Hill, 2003.
Huettenmueller, Rhonda, *Algebra Demystified*. New York, McGraw-Hill,
2003.
Krantz, Steven, *Calculus Demystified*. New York, McGraw-Hill, 2003.
Moyer, Robert and Ayres, Frank, *Trigonometry*. New York, McGraw-Hill,
1999.

Web Sites

Encyclopedia Britannica Online, www.britannica.com.
Eric Weisstein's World of Mathematics, www.mathworld.wolfram.com.

INDEX

INDEX

ABOUT THE AUTHOR

Stan Gibilisco is the author of many best-selling McGraw-Hill titles, including the *TAB Encyclopedia of Electronics for Technicians and Hobbyists, Teach Yourself Electricity and Electronics*, and *The Illustrated Dictionary of Electronics. Booklist* named his *McGraw-Hill Encyclopedia of Personal Computing* one of the "Best References of 1996."